NEURAL NETWORK PARALLEL COMPUTING

THE KLUWER INTERNATIONAL SERIES
IN ENGINEERING AND COMPUTER SCIENCE

NEURAL NETWORK PARALLEL COMPUTING

by

Yoshiyasu Takefuji
Case Western Reserve University, USA
Keio University, Japan

KLUWER ACADEMIC PUBLISHERS
Boston/Dordrecht/London

Distributors for North America:
Kluwer Academic Publishers
101 Philip Drive
Assinippi Park
Norwell, Massachusetts 02061 USA

Distributors for all other countries:
Kluwer Academic Publishers Group
Distribution Centre
Post Office Box 322
3300 AH Dordrecht, THE NETHERLANDS

Library of Congress Cataloging-in-Publication Data

Takefuji, Yoshiyasu, 1955-
Neural network parallel computing / by Yoshiyasu Takefuji.
p. cm. -- (The Kluwer international series in engineering and computer science ; SECS 0164)
Includes bibliographical references (p.) and index.
ISBN 0-7923-9190-X (acid free paper)
1. Neural networks (Computer science) 2. Parallel processing (Electronic computers) I. Title. II. Series.
QA76.87.T35 1992
006.3--dc20 91-42280
CIP

Printed on acid-free paper.

Printed in the United States of America

CONTENTS

FOREWORD REMARKS

Dr. Yoshiyasu Takefuji came to the United States from Keio University in Japan having studied under one of the famous fifth generation computer leaders, Professor Hideo Aiso. With great ambition and enthusiasm, Dr. Takefuji sought to solve all application problems by neurocomputing implemented by VLSI chips. Soon, the reality of teaching, research and funding that are all so familiar to those climbing a professor's career ladder, became a bit overwhelming to Dr. Takefuji.

Nevertheless, the author of this book kept on solving all application problems imaginable with respect to constraint satisfaction problems, by solving first order simultaneous differential equations that he called the "motion of equations" of neural networks. Dr. Takefuji's fascinating achievements are cohesively presented in Neural Network Parallel Computing.

Beginning immediately with problem solving, Dr. Takefuji has formulated the problems of N-queen, scheduling, four-coloring, planarization, routing, sequencing, tour, allocation, sorting, and tiling. Silicon chip examples and mathematical proofs are then given.

This Volume presents many useful algorithms and applications via top-down design approaches. It serves as a good reference textbook for researchers who are looking for a variety of interesting applications.

Dr. Harold Szu
(President-elect of International Neural Network Society)

Neural Network Parallel Computing by Dr. Yoshiyasu Takefuji is a veritable treasure trove for all who are interested in optimization computations with neural networks. In 1985, Hopfield and Tank first demonstrated how a fully connected recurrent net might be used to search for solutions to "truly difficult" problems of combinatorial optimization nature. Since then, there has been a long chain of related developments, many supportive of the proposal and others reporting on difficulties. Dr. Takefuji and his collaborators have re-examined a large number of optimization problems of academic and practical interest and describe in this Volume how appropriately evolved versions of the original approach can indeed yield solutions superior to benchmark solutions, estimated tediously through other routes. Dr. Takefuji brings insight and skill to this task and this Volume is well worth studying and having as a reference.

Professor Yoh-Han Pao
(Case Western Reserve University)

ACKNOWLEDGEMENTS

Since Dr. John J. Hopfield and Dr. David Tank published a paper entitled "Neural Computation of Decisions in Optimization Problems," in Biological Cybernetics in 1985, I have studied neural network computing. I would like to express my gratitude for Dr. Hopfield's pioneering work and his kindness helping my former Ph.D. student, Dr. Simon Foo now at Florida State University. I would like to thank Dean and Professor Hideo Aiso in the Environmental Information School at Keio University and Mrs. Kimiko Aiso for encouraging me to finish this book. Professor Yoh-Han Pao and his company AI WARE have always helped me study this exciting area of neural network computing. I would like to thank my former graduate students, Dr. Kuo Chun Lee, Miss Li Lin Chen, and Mr. Nobuo Funabiki for their implementations. I would like to thank Ph.D. students, Mr. C. W. Lin and Mr. Y. B. Cho. I would like to thank Dr. Robert Jonnarone at University of South Carolina who introduced me to his Conjunctoids concepts. I want to thank my former research associate, Mr. Toshimitsu Tanaka at Sumitomo Metal Industries in Japan for string search programming. I also would like to thank Dr. Harold Szu at NSWC and Dr. Yutaka Akiyama at ETL for stochastic neural computing. These neural network projects were partly supported by the National Science Foundation (MIP-8902819) and Sumitomo Metal Industries in Japan.

I would like to thank my wife Michiyo, my sons Noriyasu and Akihiro, and the forthcoming baby in December 1991. I also thank my parents Tokiyo & Tameichi Takefuji and Nobuo & Yoshio Yoshida.

NEURAL NETWORK PARALLEL COMPUTING

Chapter 1

NEURAL NETWORK MODELS AND N-QUEEN PROBLEMS

Brief history of neural network parallel computing and four mathematical neural network models including a McCulloch-Pitts neuron, a sigmoid neuron, a hysteresis McCulloch-Pitts neuron, and a maximum neuron are introduced. "What is an N-queen problem?" and a neural network model of the N-queen problem are first presented to show how to use the neural network (sequential/parallel) model for solving general optimization problems. Two important issues are highlighted in Chapter 1: I) how to construct a neural representation from a given problem, and II) how to build the motion equation considering the necessary and sufficient constraints and/or the cost function from the problem. Three Pascal programs for N-queen problems are given for the reader to readily observe the neural network system which can run on virtually any computer.

1.1 INTRODUCTION

In 1943 mathematical models based on biological computation were proposed by W. S. McCulloch and W. H. Pitts (McCulloch and Pitts 1943). They attempted to take advantage of the elegant natural biological computation in the brain of animals and human beings. D. O. Hebb presented the learning theory for realizing the associative memory where the strengths of the existing synaptic connections between neurons are modified by the input patterns (Hebb 1949). B. Widrow at Stanford University demonstrated adaptive switching circuits in 1960 (Widrow and Hoff 1960). In 1961

F. Rosenblatt at Cornell University presented Perceptrons and the theory of brain mechanisms in his book (Rosenblatt 1962). In 1969 M. Minsky and S. Papert at MIT showed the limitation of Perceptrons in their book (Minsky and Papert 1969). The negative results against the artificial neural network computing had caused less support and interest from governments/industries and consequently shrank the scale of neural network study and the number of investigators. However a small number of researchers such as S. Amari, L. Cooper, K. Fukushima, and S. Grossberg studied the neural network computing during the 1960's and 1970's. In the 1970's J. A. Anderson and T. Kohonen developed mathematical models of associative memory.

The new discovery in neurobiology and the explosive interest in parallel computation along with the inexpensive VLSI (very-large-scale-integrated) circuit technology have caused a dramatic resurgence. In 1985 J. J. Hopfield and D. Tank proposed an artificial neural network for optimization problems which has attracted many new investigators to get involved in neural computing (Hopfield and Tank 1985). However in 1988 G. V. Wilson and G. S. Pawley strongly criticized the neural network for optimization problems (Wilson and Pawley 1988) and R. Paielli reported the simulation test of the Hopfield neural network in the same problem (Paielli 1988) both of which discouraged US federal agencies to support the neural network research for optimization problems. Since 1988 it has been widely believed that the artificial neural network is not suitable for optimization problems.

This book is intended to demonstrate the capability of the artificial neural network for solving optimization problems over the best known algorithms or the best methods if they exist. The book covers a variety of professional fields including game theory, computer science, graph theory, molecular biology, VLSI computer aided design, reliability, management science, and communications and computer networks. It contains ten applications including N-queen problems, crossbar switch scheduling problems, four-coloring and k-colorability problems, graph planarization problems, channel routing problems, RNA secondary structure prediction problems, knight's tour problems, spare allocation problems, sorting and searching problems, and tiling problems. In Chapter 1 N-queen problems are first introduced for the reader to understand a basic neural network approach including how to represent a problem with artificial neurons called neural representation, how to construct the motion equation from the necessary and the sufficient constraints and/or the cost function in a given problem, and how to develop a software simulator on a Unix workstation, a

PC machine, or a Macintosh machine. Three Turbo Pascal programs are given to simulate the artificial neural network in sequential/parallel for N-queen problems in Chapter 1. The reader is encouraged to solve some of exercises in the end of every Chapter. After reading Chapter 1, Chapter 2 through 10 can be read independently except Chapter 5 and Chapter 6. In order to understand RNA secondary structure prediction problems in Chapter 6 and channel routing problems in Chapter 5, reading Chapter 4 is recommended. The important mathematical background involved in neural computing for optimization problems is summarized in Chapter 12. Chapter 13 shows on-going research applications including the module orientation problem, the maximum clique problem, the max cut problem, and other crossbar switch scheduling applications. Chapter 13 also depicts the future research of neural computing for optimization problems. Finally Chapter 14 introduces Conjunctoids and artificial learning.

1.2 MATHEMATICAL NEURAL NETWORK MODELS

The mathematical model of the artificial neural network consists of two components; neurons and synaptic links. The output signal transmitted from a neuron propagates to other neurons through the synaptic links. The state of the input signal of a neuron is determined by the linear sum of weighted input signals from the other neurons where the respective weight is the strength of the synaptic links. Every artificial neuron has the input U and the output V. The output of the *i*th neuron is given by $V_i=f(U_i)$ where f is called the neuron's input/output function. In this book we introduce a sigmoid function, a McCulloch-Pitts function, a hysteresis McCulloch-Pitts function, a modified McCulloch-Pitts function, and a maximum function as the neuron's input/output function.

The interconnections between the *i*th neuron and other neurons are determined by the motion equation. The change of the input state of the *i*th neuron is given by the partial derivatives of the computational energy function E with respect to the output of the *i*th neuron where E follows an n-variable function: $E(V_1, V_2,...,V_n)$. The motion equation of the *i*th neuron is given by:

$$\frac{dU_i}{dt} = -\frac{\partial E(V_1, V_2,...,V_n)}{\partial V_i} \quad (1.1)$$

The goal of the artificial neural network for solving optimization problems is to minimize the fabricated computational energy function E in Eq. (1.1). The energy function not only determines how many neurons should be used in the system but also the strength of the synaptic links between neurons. It is constructed by considering the necessary and sufficient constraints and/or the cost function from the given problem. It is usually easier to build the motion equation than the energy function. From Eq. (1.1) the energy function E can be obtained:

$$E=\int dE= -\int \frac{dU_i}{dt}dV_i \quad (1.2)$$

The artificial neural network provides a parallel gradient descent method to minimize the fabricated energy function E. Six convergence theorems/proofs of the artificial neural network are given in Chapter 12 to provide a mathematical background: 1) the proof of the harmfulness of using the decay term in the motion equation although it is still widely believed that the decay term is absolutely necessary, 2) the convergence theorem of the analog neural network without the decay term, 3) the convergence theorem of the McCulloch-Pitts neural network, 4) the convergence theorem of the discrete neural network, 5) the convergence theorem of the hysteresis McCulloch-Pitts neural network, and 6) the convergence theorem of the maximum neural network. It is not required for the novice reader to fully understand all of the mathematical background in Chapter 12 but it may be needed for the advanced reader.

The input/output function of the McCulloch-Pitts binary neuron model is shown in Fig. 1-1, that of the hysteresis McCulloch-Pitts neuron model in Fig. 1-2, and that of the sigmoid neuron model in Fig. 1-3 respectively.

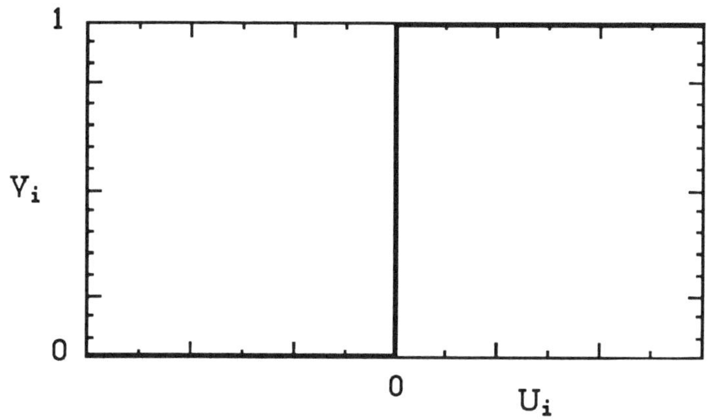

Fig. 1-1 McCulloch-Pitts Input/Output function

The McCulloch-Pitts input/output function is given by:

$$V_i=f(U_i)=1 \text{ if } U_i>0$$
$$0 \text{ otherwise} \qquad (1.3)$$

where V_i and U_i are the output and the input of the *i*th neuron respectively. If the energy function follows the quadratic form, U_i will be given by: $U_i=\sum_k W_{ki}V_k$ where W_{ki} is the strength of the synaptic link from the *k*th neuron to the *i*th neuron. The state of the McCulloch-Pitts neural network is allowed to converge to the local minimum where the convergence speed is relatively faster than that of the sigmoid neural network. However the McCulloch-Pitts neural network sometimes generates undesirable oscillatory behaviors. In order to suppress oscillatory behaviors, the hysteresis McCulloch-Pitts neuron model is introduced where the input/output function of the *i*th hysteresis neuron is given by:

$$V_i=1 \text{ if } U_i>UTP \text{ (Upper Trip Point)}$$
$$=0 \text{ if } U_i<LTP \text{ (Lower Trip Point)}$$
$$\text{unchanged otherwise} \qquad (1.4)$$

where UTP is always larger than LTP. Because of suppressing the oscillatory behavior it shortens the convergence time consequently.

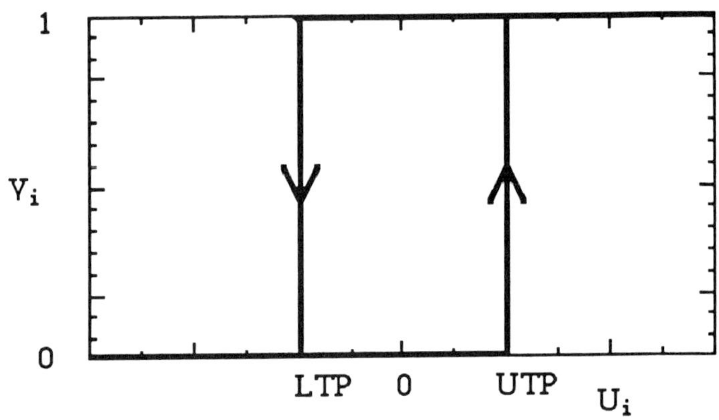

Fig.1-2 Hysteresis McCulloch-Pitts input/output function

The neural network for combinatorial optimization problems was first introduced by Hopfield and Tank in 1985 (Hopfield and Tank 1985). They use the predefined energy function E which follows the quadratic form:

$$E=\sum_{i=1}^{N}\sum_{j=1}^{N} W_{ij}V_iV_j+\sum_{i=1}^{N} V_iI_i \qquad (1.5)$$

where W_{ij} is the strength of a synaptic link between the *i*th and the *j*th neuron where the condition of $W_{ij}=W_{ji}$ must be always satisfied. Note that I_i is the constant bias of the *i*th neuron.

Hopfield gives the motion equation of the *i*th neuron (Hopfield and Tank 1985):

$$\frac{dU_i}{dt}=-\frac{U_i}{\tau}-\frac{\partial E}{\partial V_i} \qquad (1.6)$$

where the output follows the continuous, nondecreasing, and differentiable function called sigmoid function:

$$V_i=f(U_i)=\frac{1}{2}(\tanh(\lambda_0 U_i)+1) \tag{1.7}$$

where λ_0 is a constant which is called gain and it determines the slope of the sigmoid function.

Wilson and Pawley strongly criticized the Hopfield and Tank neural network through the travelling salesman problem (Wilson and Pawley 1988). Unfortunately Wilson and Pawley did not know what causes the problem. The use of the decay term ($-U_i/\tau$) in Eq. (1.6) increases the computational energy function E under some conditions instead of decreasing it. The detail of the conditions and the proof are given in Chapter 12.

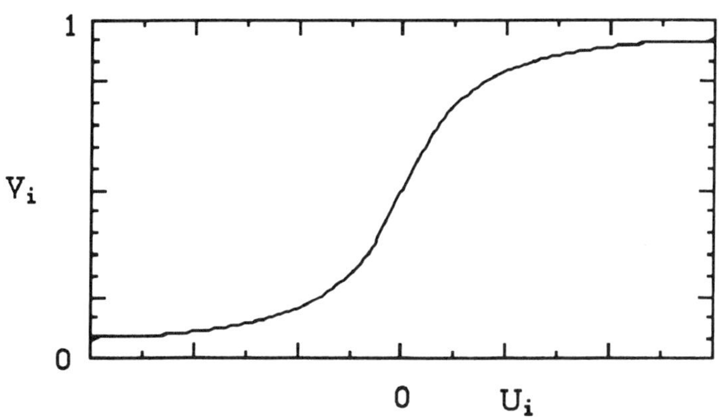

Fig. 1-3 Sigmoid input/output function

In the McCulloch-Pitts neural network with/without hysteresis or in the sigmoid neural network it is not always guaranteed that the converged state is equivalent to an acceptable solution. The acceptable solution means that it satisfies the required constraints, however it may or may not be the best solution. However in the maximum neural network it is always guaranteed to generate acceptable solutions. The maximum neural network is composed of M clusters where each cluster consists of n neurons. One and only one neuron among n neurons with the maximum input per cluster is encouraged to fire in the maximum neural network. The input/output function of the *i*th maximum neuron in the *m*th cluster is given by: $V_{m\,i}=1$ if $U_{m\,i} = \max\{ U_{m\,1}, .., U_{m\,n} \}$ and $U_{m\,i} \geq U_{m\,j}$ for i > j, and 0 otherwise. In the maximum neuron model it is always guaranteed to keep one and only one neuron to fire per cluster. The maximum neuron model has the following advantages: 1) every converged state is equivalent to a feasible solution; 2) tuning the coefficients parameters is not needed; and 3) the termination condition of the equilibrium state is given by the simple mathematical formula, while the other existing neural network models must suffer from the unacceptable solution from the converged state (where the definition of the convergence condition is mathematically unclear) and must tune the parameters in the motion equation very carefully.

1.3 N-QUEEN NEURAL NETWORK

The 8-queen problem was proposed in 1848 and it was investigated by several famous mathematicians including C. F. Gauss in 1850 where the goal of the problem is to place 8 queens on an 8 x 8 chessboard in mutually nonattacking positions. The 8-queen problem has been used as a benchmark problem to demonstrate divide-and-conquer methods (Abramson and Yung 1989), trial-and-error methods including back-tracking algorithms (Bitner and Reingold 1975) (Stone and Stone 1987), and other methods (Yaglom and Yaglom 1964) (Kale 1990). Although few parallel algorithms have been proposed (Filman and Friedman 1984) (Finkel and Manber 1987) (Abramson and Yung 1989), there exists no satisfactory parallel algorithm. Page et al. presented the first deterministic parallel algorithm based on the Hopfield neural network (Page and Tagliarini 1987). Akiyama et al. proposed an $O(N^2)$ parallel algorithm based on the stochastic neural network model (Akiyama et al. 1989). However the state of the system based on any one of the existing neural network algorithms often converges to the unacceptable local minimum. In other words, their solution quality dramatically degrades with the problem size.

The goal of a general N-queen problem is to locate N queens on an N x N chessboard where any pair of queens does not command each other. Note that a queen commands vertically, horizontally, and diagonally as shown in Fig. 1-4.

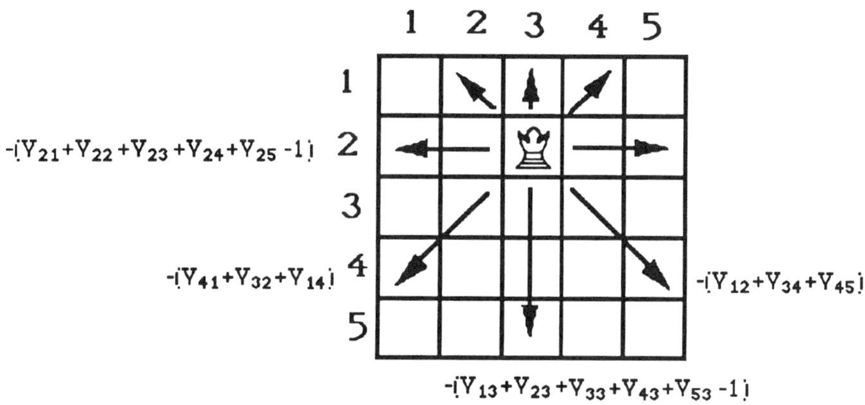

Fig. 1-4 How a queen commands

It is well understood that one and only one queen must be located per row and per column so as to locate N queens on an N x N chessboard. In other words more than one queen should not be located per row and per column. The condition of the constraints or the violations gives the interconnections of the artificial neural network. An N x N neural array is prepared for the N-queen problem where each element of the neural array represents a square of the N x N chessboard. The output state of the *ij*th neuron gives the location of a queen in the *i*th row and the *j*th column. In other words, $V_{ij}=1$ means that a queen is located in the *i*th row and the *j*th column. $V_{ij}=0$ means that no queen is located in the *i*th row and the *j*th column. For example, to locate a queen in the second row and the third column as shown in Fig. 1-4, V_{23} must be $V_{23}=1$. The motion equation of the *ij*th neuron for the N-queen problem is given by:

$$\frac{dU_{ij}}{dt} = -A\left(\sum_{k=1}^{N} V_{ik}-1\right)-A\left(\sum_{k=1}^{N} V_{kj}-1\right)$$
$$-B \sum_{\substack{1\le i-k,j-k\le N \\ k\neq 0}} V_{i-k,j-k} -B \sum_{\substack{1\le i-k,j+k\le N \\ k\neq 0}} V_{i-k,j+k} \quad \text{for } i,j=1,..,5 \qquad (1.8)$$

In Eq. (1.8) the first term gives the row constraint such that one and only one queen/neuron should be located/fired in the *i*th row. The second term describes the column constraint such that one and only one queen/neuron should be located/fired in the jth column. The third and the fourth term represent the diagonal constraints such that no pair of queens should not diagonally command each other. The following two terms must be added to Eq. (1.8) to enable the state of the system to escape from the local minimum and to converge to the global minimum or the acceptable solution:

$$+Ch\left(\sum_{k=1}^{N} V_{ik}\right) +Ch\left(\sum_{k=1}^{N} V_{kj}\right) \quad (1.9)$$

where h(x) is 1 if x=0, and 0 otherwise. The function of h(x) is called hill-climbing term. In the local minimum no neurons/queens are fired/located on the ith row or the jth column. For example in the 5-queen problem with A=B=C=1, the motion equation of the neuron in the second row and the third column as shown in Fig. 1-4 is given by:

$$\begin{aligned}\Delta U_{23} = \frac{dU_{23}}{dt} &= -(V_{21}+V_{22}+V_{23}+V_{24}+V_{25} -1) -(V_{13}+V_{23}+V_{33}+V_{43}+V_{53}-1) \\ &-(V_{12}+V_{34}+V_{45}) -(V_{14}+V_{32}+V_{41}) \\ &+h(V_{21}+V_{22}+V_{23}+V_{24}+V_{25})+h(V_{13}+V_{23}+V_{33}+V_{43}+V_{53}) \quad (1.10)\end{aligned}$$

Fig. 1-5 and Fig. 1-6 show one of the solutions for the 5-queen problem and one of the local minima for the 6-queen problem respectively. Remember that in the local minima less than N queens are placed on an N x N chessboard but any additional queen cannot be placed in any row or column. For example, as shown in Fig. 1-6 we cannot locate any queen in the second column or in the fourth row so that we say, the state of the system is in the local minimum.

Fig. 1-5 One of the solutions

Fig. 1-6 One of the local minima

1.4 GENERAL OPTIMIZATION PROGRAMS

Since a parallel machine is not widely available to many readers for parallel simulation, two programs are provided to simulate the behavior of the parallel machine using a sequential machine as if the program is running on the parallel machine. In order to distinguish a sequential model and two parallel models, three general programs for solving optimization problems are given: Program1 to simulate a sequential neural network model, Program2 for a synchronous parallel neural network model using a sequential machine, and Program 3 for an asynchronous parallel neural network model using a sequential machine.

In order to numerically solve the partial differential equation or the differential equation, the first order Euler method is widely used where it is the simplest among the existing numerical methods. Based on the first order Euler method, the value of U(t+1) is determined by U(t) and ΔU(t):

$$U(t+1)=U(t)+\Delta U(t)\cdot \Delta t \qquad (1.11)$$

where ΔU(t) is given by Eq. (1.1). With Δt=1, Eq. (1.11) for N-queen problems (where a two dimensional neural array is used) is given by:

$$U_{ij}(t+1)=U_{ij}(t)+\Delta U_{ij}(t) \qquad \text{for } i,j=1,\ldots,N \qquad (1.12)$$

In order to solve N-queen problems, the initial values of $U_{ij}(0)$ for i,j=1,...,N are randomly given and Eq. (1.12) is iteratively used until the state of the system reaches the equilibrium state. However the equilibrium state is not always equivalent to the acceptable solution. In the N-queen problem, the condition of the program termination is given by:

$$\Delta U_{ij}(t)=0 \qquad \text{for } i,j=1,\ldots,N \qquad (1.13)$$

The condition of Eq. (1.13) implies that the constraints are all satisfied. In other words, no violations can be found on the condition. Eq. (1.12) is either sequentially updated or parallelly updated. In parallel computing there are a synchronous parallel model and an asynchronous parallel model. In the synchronous parallel model, clock signals or handshaking signals are used for synchronous communications between processing elements (neurons). In the asynchronous parallel model the state of every processing element (neuron) is updated asynchronously without using any clock/handshaking signals. In Fig. 1-7 Program1 shows a general program for simulating the sequential neural network model where a two dimensional neural representation and the McCulloch-Pitts neuron model are used. Of course, it is possible to use a single dimensional neural representation or a higher dimensional neural representation in Program1. As soon as updating the value of U_{ij}, the value of V_{ij} is evaluated based on the input/output function of the neuron model in Program1.

It is quite straightforward to simulate a parallel computing model using a sequential machine. Of course if the reader is able to use a parallel machine the motion equations can be solved parallelly on it. There are two parallel computing models: a synchronous parallel model and an asynchronous parallel model. Program2 in Fig. 1-8 illustrates a program to simulate the synchronous parallel model which runs on a sequential machine as if it runs on a parallel machine. In the first loop all input values of U_{ij} for i,j=1,...,N are sequentially updated while all output values of V_{ij} for i,j=1,...,N are fixed. Then in the second loop all output values of V_{ij} for i,j=1,...,N are sequentially updated while all input values of U_{ij} for i,j=1,...,N are fixed. It is equivalent to simultaneously updating the values of all inputs and outputs. In Program2 the McCulloch-Pitts neuron model is used:

```
Program1 sequential-simulator
begin
initialization of Uij and Vij for i,j:=1 to N;
/*** Main Program ***/
 while (a set of conflicts is not empty) do
  for i:=1 to N
  for j:=1 to N
      begin
            Uij:=Uij+ ΔUij;
            If Uij>0 then Vij:=1 else Vij:=0;
      end;

end;
/*** Main Program end ***/
```

Fig. 1-7 Simulator for a sequential neural network model

Program3 in Fig. 1-9 shows the asynchronous parallel simulator to run on a sequential machine. Program2 is very similar to Program3 except the statement of "If random<x then." The variable "random" is an integer with a range from 0 to y where x is the constant number and x satisfies 0<x<y. If random is smaller than x then update the state of the output, otherwise keep the previous state of the output. The statement of "If random<x then" provides the pseudo asynchronous system behavior in simulation.

```
Program2 synchronous-parallel-simulator
begin
initialization of Uij and Vij for i,j:=1 to N;
/*** Main Program ***/
 while (a set of conflicts is not empty) do
 begin
/*** The first loop ***/
   for i:=1 to N
   for j:=1 to N
     Uij:=Uij+ΔUij;
/*** End of the first loop ***/
/*** The second loop ***/
   for i:=1 to N
   for j:=1 to N
     If Uij>0 then Vij:=1 else Vij:=0;
/*** End of the second loop ***/
 end;
end;
/*** Main Program end ***/
```

Fig. 1-8 Synchronous parallel simulator on a sequential machine

```
Program3 Asynchronous-parallel-simulator
begin
initialization of Uij and Vij for i,j:=1 to N;
/*** Main Program ***/
 while (a set of conflicts is not empty) do
 begin
/*** The first loop ***/
   for i:=1 to N
   for j:=1 to N
    Uij:=Uij+ΔUij;
/*** End of the first loop ***/
/*** The second loop ***/
   for i:=1 to N
   for j:=1 to N
    If random<x then
     If Uij>0 then Vij:=1 else Vij:=0;
/*** End of the second loop ***/
 end;
end;
/*** Main Program end ***/
```

Fig. 1-9 Asynchronous parallel simulator on a sequential machine

1.5 N-QUEEN SIMULATION PROGRAMS

Fig. 1-10 shows a Turbo Pascal program which is based on Program1 as shown in Fig. 1-7. It is ready to run on any Macintosh machine. The program is to simulate the sequential N-queen neural network model. In the program the first term and the second term of the motion equation in Eq. (1.8) are realized by:

```
        sum_column:=0;sum_row:=0;
        for k:=1 to max do begin sum_row:=sum_row+V[i,k];
sum_column:=sum_column+V[k,j]; end;
```

where max is the N-queen problem size. The third term and the fourth term in Eq. (1.8) or two diagonal constraints are given by:

```
        diagonal1:=0; k:=1;
       while((j+k)<=max) and ((i-k)>=1) do begin diagonal1:=diagonal1+V[i-
k,j+k]; k:=k+1; end;
        k:=1;
```

```
        while((j-k)>=1) and ((i+k)<=max) do begin diagonal1:=diagonal1+V[i+k,j-
k]; k:=k+1; end;
         diagonal2:=0; k:=1;
        while((j+k)<=max) and ((i+k)<=max) do begin
diagonal2:=diagonal2+V[i+k,j+k]; k:=k+1; end;
         k:=1;
        while((j-k)>=1) and ((i-k)>=1) do begin diagonal2:=diagonal2+V[i-k,j-k];
k:=k+1; end;
```

The hill-climbing term in Eq. (1.9) is given by:

```
h:=0;
if (sum_column=0) then h:=1; if (sum_row=0) then h:=h+1;
```

Therefore the new input state of the *ij*th neuron in Eq. (1.12) is finally computed by:

```
U[i,j]:=U[i,j]-A*(sum_row+sum_column-2)-B*(diagonal1+diagonal2)+C*h;
```

In order to shorten the convergence time, the values of the inputs are always kept in the certain range. The value of the input U_{ij}, the output value of the *ij*th neuron is evaluated by:

```
if (U[i,j]>15) then U[i,j]:=15; if (U[i,j]<-5) then U[i,j]:=-5;
            if U[i,j]>0 then V[i,j]:=1 else V[i,j]:=0;
```

where the maximum and the minimum value of the input are given by +15 and -5 respectively. The termination condition in Eq. (1.13) is realized by:

```
diag:=1;
  while ((not keypressed) and (diag>0) and (t<2000)) do
   begin
    diag:=0;
    for i:=1 to max do
     for j:=1 to max do
      begin
       conf:=1;
        if((sum_column+sum_row=2) and (diagonal1<2) and (diagonal2<2)) then conf:=0;
       diag:=diag+conf;
      end;{******** The end of i and j loop ********}
 end; {******** The end of while loop ********}
```

If diag=0 then the program will be automatically terminated. The following routine displays the simulation result of the N-queen problem on the Macintosh screen.

```
clearscreen;
 for i:=1 to max do
  for j:=1 to max do
   if (V[i,j]=1) then
     begin
      SetRect(tempRect,0,0,length,length); OffsetRect(tempRect,(i-1)*length,(j-
1)*length); PaintRect(tempRect);
     end;
   MoveTo(0,0); LineTo(0,max*length); MoveTo(0,max*length);
LineTo(max*length,max*length);
   MoveTo(max*length,max*length); LineTo(max*length,0);
MoveTo(max*length,0); LineTo(0,0);
 for i:=1 to max do
   begin
    MoveTo(i*length,0); LineTo(i*length,max*length); MoveTo(0,i*length);
LineTo(max*length,i*length);
   end;
gotoXY(50,1);writeln('No. of steps=',t);
```

Fig. 1-11 shows a Turbo Pascal program for simulating the synchronous parallel N-queen neural network model where the hysteresis McCulloch-Pitts neuron is used. UTP=3 and LTP=0 are used for hysteresis.

Fig. 1-12 depicts a Turbo Pascal program for simulating the asynchronous parallel N-queen neural network model where the same hysteresis McCulloch-Pitts neuron model is used. The difference between the sequential-simulation program in Fig. 1-10 and the synchronous-parallel-simulation program in Fig. 1-11 or the asynchronous-parallel simulation program in Fig. 1-12 lies in that the output values of the neurons are simultaneously updated outside of the motion equation loop for parallel simulation, while the output value of every neuron is individually computed as soon as the input of the neuron is evaluated inside of the motion equation loop for sequential simulation.

The asynchronous parallel system has a distinguished advantage over the synchronous parallel system. The asynchronous parallel system does not require any clock or synchronization mechanism. In Fig. 1-12 the program attempts to sequentially simulate the asynchronous parallel N-queen neural network model. The statement of "if (abs(random) mod 2 <1) then" in the output evaluation routine provides the pseudo asynchronous behavior in simulation.

Fig. 1-13 depicts several simulation results of N-queen problems using the illustrated programs in Fig. 1-10, Fig. 1-11, and Fig. 1-12. It takes about 11 minutes to solve a 100-queen problem on a Macintosh SE/30 where it only requires 36 iteration steps. If 10,000 processing elements are provided and assume that each processing element has the processing speed of one million state-updates per second (1 MSUPS), it will take 36 micro seconds to solve the 100-queen problem.

```
program sequential_queen;
uses memtypes, quickdraw,osintf,ToolIntf;
type twod_int= array[1..89,1..89] of integer;
var
A,B,C,t,h,i,j,k,sum_column,sum_row,diagonal1,diagonal2,conf,diag,seed,row,length
,max:integer;
U,V:twod_int; tempRect : Rect;
begin
   A:=1; B:=1; C:=1;
   writeln('Please define the queen problem size (6-89).'); readln(max);
   writeln('Please input a seed(0-9999).'); readln(seed);
   for i:=1 to seed do U[1,1]:=random; length:=trunc(270/max);
   for i:=1 to max do
     for j:=1 to max do
     begin U[i,j]:=-trunc(abs(random/4000)); if U[i,j]>0 then V[i,j]:=1 else V[i,j]:=0;
end;
{******** Main program ********}
   t:=0; diag:=1;
   while ((not keypressed) and (diag>0) and (t<2000)) do
     begin diag:=0;
       for i:=1 to max do
        for j:=1 to max do
          begin
           sum_column:=0;sum_row:=0;
           for k:=1 to max do begin sum_row:=sum_row+V[i,k];
           sum_column:=sum_column+V[k,j]; end;
           diagonal1:=0; k:=1;
          while((j+k)<=max) and ((i-k)>=1)
          do begin diagonal1:=diagonal1+V[i-k,j+k]; k:=k+1; end;
           k:=1;
          while((j-k)>=1) and ((i+k)<=max)
          do begin diagonal1:=diagonal1+V[i+k,j-k]; k:=k+1; end;
           diagonal2:=0; k:=1;
          while((j+k)<=max) and ((i+k)<=max)
          do begin diagonal2:=diagonal2+V[i+k,j+k]; k:=k+1; end;
           k:=1;
          while((j-k)>=1) and ((i-k)>=1) do begin diagonal2:=diagonal2+V[i-k,j-k];
           k:=k+1; end;
          h:=0;conf:=1;
```

```
      if (sum_column=0) then h:=1; if (sum_row=0) then h:=h+1;
     if((sum_column+sum_row=2) and (diagonal1<2) and (diagonal2<2)) then
      conf:=0;
     U[i,j]:=U[i,j]-A*(sum_row+sum_column-2)-B*(diagonal1+diagonal2)+C*h;
      if (U[i,j]>15) then U[i,j]:=15; if (U[i,j]<-5) then U[i,j]:=-5;
         if U[i,j]>0 then V[i,j]:=1 else V[i,j]:=0;
     diag:=diag+conf;
   end;{******** The end of i and j loop ********}
    t:=t+1; if (t mod 20 <5) then C:=4 else C:=1; write(t);
 end; {******** The end of while loop ********}
clearscreen;
for i:=1 to max do
 for j:=1 to max do
  if (V[i,j]=1) then
    begin
     SetRect(tempRect,0,0,length,length); OffsetRect(tempRect,(i-1)*length,(j-
1)*length); PaintRect(tempRect);
    end;
  MoveTo(0,0); LineTo(0,max*length); MoveTo(0,max*length);
LineTo(max*length,max*length);
  MoveTo(max*length,max*length); LineTo(max*length,0);
MoveTo(max*length,0); LineTo(0,0);
 for i:=1 to max do
   begin
    MoveTo(i*length,0); LineTo(i*length,max*length); MoveTo(0,i*length);
LineTo(max*length,i*length);
   end;
gotoXY(50,1);writeln('No. of steps=',t);readln;readln;
end. {******** The end of program ********}
```

Fig. 1-10 A program of the sequential N-queen neural network

```
program sync_parallel_queen;
uses memtypes, quickdraw,osintf,ToolIntf;
type twod_int= array[1..89,1..89] of integer;
var
A,B,C,t,h,i,j,k,sum_column,sum_row,diagonal1,diagonal2,conf,diag,seed,row,length
,max:integer;
U,V:twod_int; tempRect : Rect;
begin
   A:=1; B:=1; C:=1;
   writeln('Please define the queen problem size (6-89).'); readln(max);
   writeln('Please input a seed(0-9999).'); readln(seed);
   for i:=1 to seed do U[1,1]:=random; length:=trunc(270/max);
   for i:=1 to max do
     for j:=1 to max do
```

```
    begin U[i,j]:=-trunc(abs(random/1600)); if U[i,j]>0 then V[i,j]:=1 else V[i,j]:=0;
end;
{******** Main program ********}
   t:=0; diag:=1;
   while ((not keypressed) and (diag>0) and (t<2000)) do
     begin diag:=0;
       for i:=1 to max do
        for j:=1 to max do
         begin
           sum_column:=0;sum_row:=0;
          for k:=1 to max do begin sum_row:=sum_row+V[i,k];
sum_column:=sum_column+V[k,j]; end;
          diagonal1:=0; k:=1;
         while((j+k)<=max) and ((i-k)>=1) do begin diagonal1:=diagonal1+V[i-
k,j+k]; k:=k+1; end;
          k:=1;
         while((j-k)>=1) and ((i+k)<=max) do begin diagonal1:=diagonal1+V[i+k,j-
k]; k:=k+1; end;
          diagonal2:=0; k:=1;
         while((j+k)<=max) and ((i+k)<=max) do begin
diagonal2:=diagonal2+V[i+k,j+k]; k:=k+1; end;
          k:=1;
         while((j-k)>=1) and ((i-k)>=1) do begin diagonal2:=diagonal2+V[i-k,j-k];
k:=k+1; end;
      h:=0;conf:=1;
      if (sum_column=0) then h:=1; if (sum_row=0) then h:=h+1;
     if((sum_column+sum_row=2) and (diagonal1<2) and (diagonal2<2)) then
conf:=0;
     U[i,j]:=U[i,j]-A*(sum_row+sum_column-2)-B*(diagonal1+diagonal2)+C*h;;
      if (U[i,j]>15) then U[i,j]:=15; if (U[i,j]<-20) then U[i,j]:=-20;
      diag:=diag+conf;
    end;{******** The end of i and j loop ********}
for i:=1 to max do
     for j:=1 to max do
     begin if U[i,j]>3 then V[i,j]:=1; if U[i,j]<0 then V[i,j]:=0; end;
      t:=t+1; if (t mod 20 <5) then C:=4 else C:=1; write(t);
 end; {******** The end of while loop ********}
clearscreen;
for i:=1 to max do
 for j:=1 to max do
  if (V[i,j]=1) then
    begin
      SetRect(tempRect,0,0,length,length); OffsetRect(tempRect,(i-1)*length,(j-
1)*length); PaintRect(tempRect);
    end;
  MoveTo(0,0); LineTo(0,max*length); MoveTo(0,max*length);
LineTo(max*length,max*length);
  MoveTo(max*length,max*length); LineTo(max*length,0);
MoveTo(max*length,0); LineTo(0,0);
```

```
for i:=1 to max do
  begin
    MoveTo(i*length,0); LineTo(i*length,max*length); MoveTo(0,i*length);
LineTo(max*length,i*length);
  end;
gotoXY(50,1);writeln('No. of steps=',t);readln;readln;
end. {******** The end of program ********}
```

Fig. 1-11 A program of the synchronous-parallel N-queen neural network

```
program async_parallel_queen;
uses memtypes, quickdraw,osintf,ToolIntf;
type twod_int= array[1..89,1..89] of integer;
var
A,B,C,t,h,i,j,k,sum_column,sum_row,diagonal1,diagonal2,conf,diag,seed,row,length
,max:integer;
U,V:twod_int; tempRect : Rect;
begin
   A:=1; B:=1; C:=1;
   writeln('Please define the queen problem size (6-89).'); readln(max);
   writeln('Please input a seed(0-9999).'); readln(seed);
   for i:=1 to seed do U[1,1]:=random; length:=trunc(270/max);
   for i:=1 to max do
     for j:=1 to max do
     begin U[i,j]:=-trunc(abs(random/1600)); if U[i,j]>0 then V[i,j]:=1 else V[i,j]:=0;
end;
{******** Main program ********}
   t:=0; diag:=1;
   while ((not keypressed) and (diag>0) and (t<2000)) do
     begin diag:=0;
       for i:=1 to max do
        for j:=1 to max do
         begin
          sum_column:=0;sum_row:=0;
          for k:=1 to max do begin sum_row:=sum_row+V[i,k];
sum_column:=sum_column+V[k,j]; end;
          diagonal1:=0; k:=1;
         while((j+k)<=max) and ((i-k)>=1) do begin diagonal1:=diagonal1+V[i-
k,j+k]; k:=k+1; end;
          k:=1;
         while((j-k)>=1) and ((i+k)<=max) do begin diagonal1:=diagonal1+V[i+k,j-
k]; k:=k+1; end;
          diagonal2:=0; k:=1;
          while((j+k)<=max) and ((i+k)<=max) do begin
diagonal2:=diagonal2+V[i+k,j+k]; k:=k+1; end;
          k:=1;
         while((j-k)>=1) and ((i-k)>=1) do begin diagonal2:=diagonal2+V[i-k,j-k];
```

```
k:=k+1; end;
      h:=0;conf:=1;
      if (sum_column=0) then h:=1; if (sum_row=0) then h:=h+1;
     if((sum_column+sum_row=2) and (diagonal1<2) and (diagonal2<2)) then
conf:=0;
     U[i,j]:=U[i,j]-A*(sum_row+sum_column-2)-B*(diagonal1+diagonal2)+C*h;;
      if (U[i,j]>15) then U[i,j]:=15; if (U[i,j]<-20) then U[i,j]:=-20;
      diag:=diag+conf;
    end;{******** The end of i and j loop ********}
for i:=1 to max do
    for j:=1 to max do if (abs(random) mod 2 <1) then
    begin if U[i,j]>3 then V[i,j]:=1; if U[i,j]<0 then V[i,j]:=0; end;
     t:=t+1; if (t mod 20 <5) then C:=4 else C:=1; write(t);
 end; {******** The end of while loop ********}
 clearscreen;
 for i:=1 to max do
  for j:=1 to max do
   if (V[i,j]=1) then
     begin
      SetRect(tempRect,0,0,length,length); OffsetRect(tempRect,(i-1)*length,(j-
1)*length); PaintRect(tempRect);
     end;
   MoveTo(0,0); LineTo(0,max*length); MoveTo(0,max*length);
LineTo(max*length,max*length);
   MoveTo(max*length,max*length); LineTo(max*length,0);
MoveTo(max*length,0); LineTo(0,0);
 for i:=1 to max do
   begin
     MoveTo(i*length,0); LineTo(i*length,max*length); MoveTo(0,i*length);
LineTo(max*length,i*length);
   end;
gotoXY(50,1);writeln('No. of steps=',t);readln;readln;
end. {******** The end of program ********}
```

Fig. 1-12 A program for simulating the asynchronous-parallel N-queen neural network model

1.6 REFERENCES

Abramson B., and Yung M., (1989), Divide and conquer under global constraints: a solution to the N-queens problem, J. of Parallel and Distributed Computing, 6, 649-662.

Akiyama Y., Yamashita A., Kajiura M., and Aiso H., (1989), Combinatorial optimization with Gaussian machines. Proc. of the Int. Joint Conf. on Neural Networks.

Bitner J, and Reingold E. M., (1975), Backtrack programming techniques, Comm. ACM 18, 651-655.

Filman R. E., and Friedman D. P., (1984), Coordinated computing: tools and techniques for distributed software, (McGraw-Hill, New York).

Finkel R., and Manber U., (1987), DIB-a distributed implementation of backtracking, ACM Trans. Programming Language Systems, 9, 235-256.

Hebb D. O, (1949), The organization of behavior (New York: Wiley).

Hopfield J. J., and Tank D. W., (1985), Neural Computation of Decisions in Optimization Problems. Biological Cybernetics, 52, 141-152.

Kale L.V., (1990), An almost perfect heuristic for the N nonattacking queens problem, Information Processing Letters, 34, 173-178.

McCulloch W. S., and Pitts W. H., (1943), A logical calculus of ideas immanent in nervous activity. Bulletin of Mathematical Biophysics, 5, 115.

Minsky M, and Papert S., (1969), Perceptrons (MIT Press).

Page E., and Tagliarini G. A, (1987), Solving constraint satisfaction problems with neural networks, Proc. of IEEE first Int. Conf. on Neural Networks.

Paielli R. A., (1988), Simulation tests of the optimization method of Hopfield and Tank using neural networks, NASA Technical Memorandom 101047.

Rosenblatt F., (1962), Principles of neurodynamics (New York: Spartan).

Stone H., and Stone J. M., (1987), Efficient search techniques-an empirical study of N-queens problem, IBM J. Res. Develop., 31, 4.

Widrow B, and Hoff M. E., (Sept. 1960), Adaptive switching circuits. Proc. of IREWESCON Convention Record.

Wilson G. V. and Pawley G. S., (1988), On stability of the Travelling Salesman Problem Algorithm of Hopfield and Tank. Bio. Cybern., 58, 63-70.

Yaglom A. M., and Yaglom I. M., (1964), Challenging mathematical problems with elementary solutions, (Holden-Day, San Francisco).

1.7 EXERCISES

1. Simulate the 5-queen neural network using the sequential program in Fig. 1-10 in order to observe whether the system has the local minimum or not. If yes, give an example of the local minima. Create some of the local minima in the 8-queen neural network using the program in Fig. 1-10.
2. Simulate the 8-queen neural network using the sequential, the synchronous parallel, and the asynchronous parallel simulation program. Collect the data on the frequency of failures and on the average number of iteration steps to converge to solutions in the successful cases. Investigate the behavior of every system with or without hysteresis, and with or without the hill-climbing term. Observe the effect of hysteresis by changing the value of UTP and LTP respectively. Explain a role of the statement of "if (t mod 20 <5) then C:=4 else C:=1;" in every program in Fig. 1-10, Fig. 1-11, and Fig. 1-12. Observe the effect of the hill-climbing term by changing the value of the coefficient C and the period in the motion equation.
3. Why is the statement of "if (U[i,j]>15) then U[i,j]:=15; if (U[i,j]<-20) then U[i,j]:=-20;" used in the program in Fig. 1-11 and Fig. 1-12? Remove the similar statement from the program in Fig. 1-10 and investigate the behavior of the system in terms of the number of iteration steps.
4. Plot the relationship between the problem size and the number of average iteration steps where at least 100 simulation runs must be performed in every problem. The range of the problem size is from 6 to 80.
5. Compare some of the conventional algorithms with the neural network algorithm for N-queen problems in terms of the computation time and the system space.
6. On an 8 x 8 chessboard we want to command 64 squares by using the minimum number of queens.

6.1 Give the neural representation and the motion equation.

6.2 Write a program based on the result of 6.1.

6.3 Collect the data on the frequency of failures and the average number of iteration steps to converge to the solution.

7. Examine the same problem using the minimum number of knights on an 8 x 8 chessboard.

8. Write a program for solving the N-superqueen problem where a solution of the N-queen problem gives a solution of the (N-1)-queen problem by removing the top row and leftmost column from the N x N chessboard. This leaves an (N-1) x (N-1) chessboard with (N-1) mutually nonattacking queens.

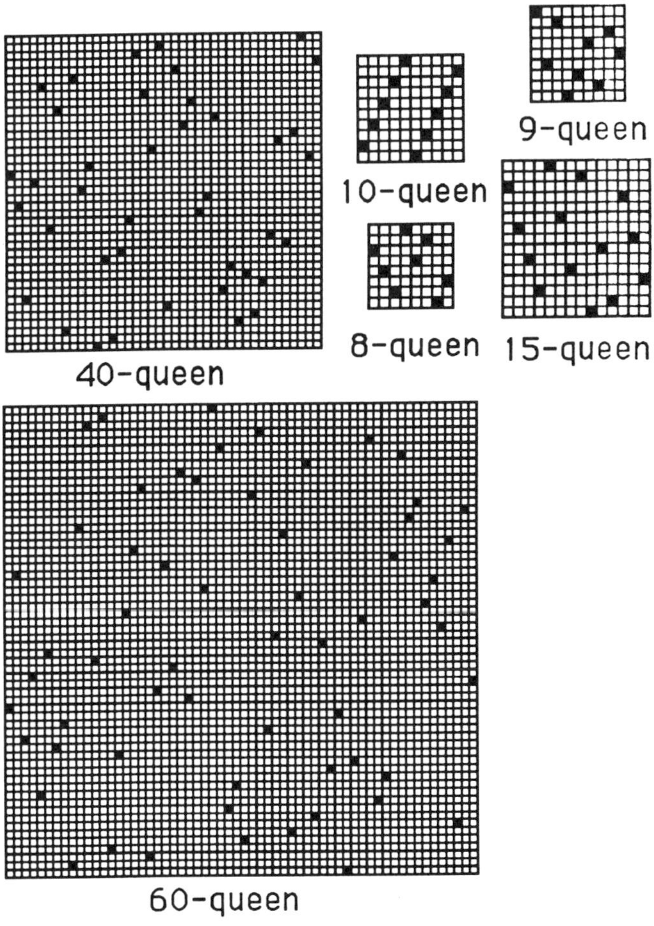

Fig. 1-13 (a) Solutions of N-queen problems

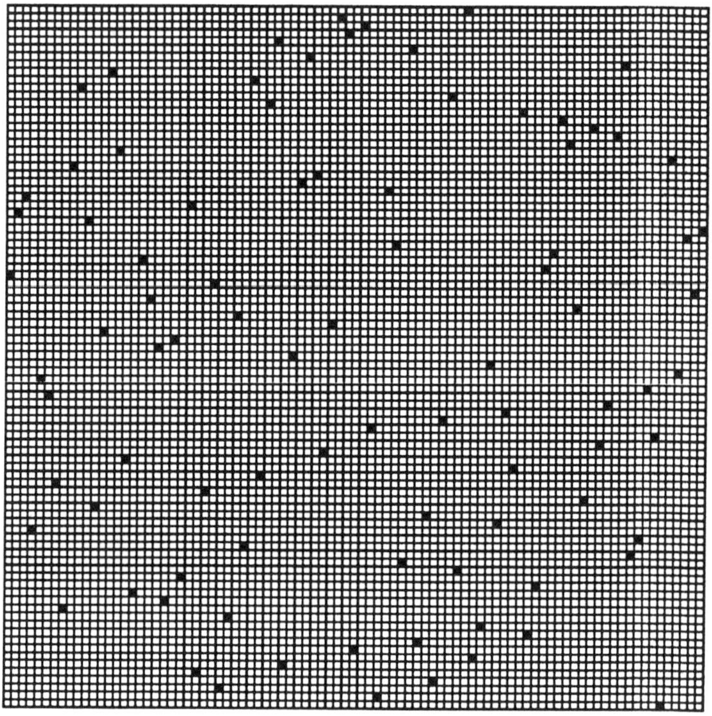

Fig. 1-13 (b) A solution of 89-queen problems

Chapter 2

CROSSBAR SWITCH SCHEDULING PROBLEMS

A hysteresis McCulloch-Pitts neuron model is used in order to suppress the complicated oscillatory behavior of neural dynamics. The artificial hysteresis McCulloch-Pitts binary neural network is used for scheduling time-multiplex crossbar switches in order to demonstrate the effects of hysteresis. Time-multiplex crossbar switching systems must control traffic on demand such that packet blocking probability and packet waiting time are minimized. The system using n x n processing elements (neurons) solves an n x n crossbar-control problem within nearly O(1) time, while the best existing parallel algorithm requires O(n) time. The hysteresis McCulloch-Pitts binary neural network maximizes the throughput of packets through a crossbar switch. The solution quality of our system does not degrade with the problem size as far as we have observed the system behavior. The relationship between N-queen problems and crossbar switch scheduling problems is also given. This Chapter is largely based on a paper published in Biological Cybernetics (Takefuji and Lee 1991).

2.1 INTRODUCTION

In time-multiplex communication systems, crossbar packet switches route traffic from the input to output where a message packet is transmitted from the source to the destination. The randomly incoming traffic must be controlled and scheduled to eliminate conflict at the crossbar switch where the conflict is that two or more users may simultaneously access to a single output. The goal of the traffic-scheduling for time-multiplex crossbar switches is not only to maximize the throughput of packets through a crossbar switch but also to minimize packet blocking probability and packet waiting time.

A request for packet transmission through an n x n crossbar is described by an n x n traffic matrix T. In the traffic matrix T, each element t_{ij} represents a request of packets from input i to output j. For example, $t_{ij}=0$ means that there is no packet to be transmitted on the jth output line from the ith input line. $t_{ij}=1$ means that at least one packet on the *i*th input line should be transmitted on the *j*th output line of the crossbar. Fig. 2-1a shows a 5 x 5 crossbar switch where 7 packets at least are requested: $t_{11}=t_{22}=t_{24}=t_{35}=t_{42}=t_{45}=t_{53}=1$. 5 packets are maximally transmitted simultaneously through the 5 x 5 crossbar switch. Fig. 2-1b depicts the solution of the problem in Fig. 2-1a where two requests t_{22} and t_{45} are rejected.

Fig. 2-1 (a) 5 x 5 crossbar switch problem

Fig. 2-1 (b) One of the solutions

In 1979 Inukai at COMSAT lab. proposed the $O(n^2)$ sequential algorithm for the n x n crossbar switch problem (Inukai 1979). In 1989 Rose at AT&T Bell lab. presented the O(n) parallel algorithm based on a cellular automaton where n^2 processing elements are used for solving an n x n traffic matrix problem (Rose 1989). Chen, Mavor, Denyer, and Renshaw proposed the $O(n^2)$ sequential algorithm of traffic routing problems for the multiprocessor system (Chen et al. 1990). They proved that the problem is NP-complete (Chen et al. 1990). In 1989 Marrakchi and Troudet at Bellcore proposed the n x n neural network algorithm based on Hopfield network model (Marrakchi and Troudet 1989). However with the Hopfield neural network, the state of the system is forced to converge to the local minimum. In other words, the solution quality drastically degrades with the problem size. Takefuji and Lee have successfully used the Hopfield neural network with McCulloch-Pitts binary neurons for solving the graph planarization problem (Takefuji and Lee 1989) and the tiling problem (Takefuji and Lee 1990a) where the state of the system converges to the near-global minimum in O(1) time. They proved that the state of the binary neural network system is guaranteed to converge to the local minimum (Takefuji and Lee 1990b). In 1986 Hoffman and Benson introduced sigmoid neurons with hysteresis for learning, where any changes in synaptic connection strengths are replaced by hysteresis (Hoffman and Benson 1986). Due to the hysteresis associated with each neuron, the system tends to stay in the region of phase space where it is located. They proposed the theory on a role for sleep in learning (Hoffman and Benson 1986).

Dynamic and static hysteresis in Crayfish stretch receptors was reported by Segundo and Martinez in 1985 (Segundo and Martinez 1985). They stated that hysteresis may be more widespread than suspected in sensory and perhaps other system. In 1989 Keeler, Pichler, and Ross presented the effects of hysteresis in pattern recognition and learning for improving the signal-to-noise ratio (Keeler et al. 1989). In this Chapter the hysteresis property is exploited in order to reduce the complicated oscillatory behaviors of neural dynamics for solving combinatorial optimization problems. The hysteresis with each neuron enhances the state of the system to stay in the region of phase space where it is located. In other words, it suppresses the oscillatory behaviors of neural dynamics so that the convergence time to the global minimum is drastically shortened.

The McCulloch-Pitts neuron model is a binary unit whose value depends on the linear sum of weighted inputs from the other neurons in the network (McCulloch and Pitts 1943). Fig. 1-1 in Chapter 1 shows the input/output relation of the McCulloch-Pitts neuron model. In 1982 Hopfield proposed the continuous input/output unit called the sigmoid neuron model (Hopfield 1982) as shown in Fig. 1-3. Simic presented the molecular electronic device with hysteresis in 1986 where it has the sigmoid hysteresis (Simic 1986).

In this Chapter a McCulloch-Pitts binary neuron model with hysteresis is highlighted. Fig. 1-2 shows the input/output function of the hysteresis McCulloch-

Pitts binary neuron model. A binary neural network with hysteresis is used for scheduling time-multiplex crossbar switches in order to demonstrate the effects of hysteresis. The complicated oscillatory behavior is one of the most undesirable phenomena in neural dynamics for solving optimization problems where we lack the mathematical tools to manipulate and understand them at a computational level (Hopfield and Tank 1986). Hysteresis suppresses the oscillatory behaviors of neural dynamics and consequently it shortens the convergence time to the global minimum. In other words, hysteresis in individual neurons allows the state of the proposed neural network to converge to the global minimum in nearly O(1) time. The system uses an N x N neural network array for solving an n x n crossbar switch problem where the output of the ijth hysteresis neuron V_{ij} is given by: $V_{ij}=1$ if $U_{ij}>UTP$ (upper trip point), 0 if $U_{ij}<LTP$ (lower trip point), and unchanged otherwise. Note that U_{ij} is the input of the *ij*th neuron. The output at any particular time does not depend only upon the present value of the input but also upon past values.

The proposed hysteresis McCulloch-Pitts binary neural network not only maximizes the throughput of packets through a crossbar switch but also minimizes packet blocking probability and packet waiting time. The constraints on an N x N crossbar switch are that no two inputs may be connected to the same output simultaneously and that no one input may be connected to more than one output simultaneously. In other words, no two packets should share the same row and the column of the traffic matrix.

The system simulator was developed based on the proposed model. A large number of simulation results are shown and demonstrated in order to support the effects of hysteresis.

2.2 CROSSBAR PROBLEMS AND N-QUEEN PROBLEMS

The constraints are considered that no two packets should share the same row and the column of the N x N traffic matrix. Our system uses an N x N neural network array where the motion equation of the *ij*th neuron is given by:

$$\frac{dU_{ij}}{dt} = -A\left(\sum_{k=1}^{N} V_{ik}-1\right) -A\left(\sum_{k=1}^{N} V_{kj}-1\right) +Bh\left(\sum_{k=1}^{N} V_{ik}\right) +Bh\left(\sum_{k=1}^{N} V_{kj}\right) \tag{2.1}$$

where h(x) is called the hill-climbing term, h(x) is 1 if x=0, 0 otherwise. Note that coefficients A and B are constant integers. The first term and the second term are the row constraint and the column constraint respectively. The first term forces one and only one neuron to be fired per row. If no neuron is fired per row or per column then it will perform excitatory forces. If more than one neuron are fired per row or per column then it will act inhibitory forces. The third term and the last term are the row hill-climbing term and the column hill-climbing term respectively. The hill-climbing terms are activated only when local conflicts (where no neuron is fired per row or column) are detected. If there is no conflict the hill-climbing terms will perform no operation. In other words, the local conflicts are resolved by the hill-climbing terms where the *ij*th neuron is encouraged to fire if the neuron has conflicts.

The reader might associate with the N-queen problem from the crossbar switch problem since they are very similar to each other except the diagonal constraints in N-queen problems. Eq. (1.8) with the hill-climbing term in Eq. (1.9) is exactly equivalent to Eq. (2.1) after removing B-term (diagonal terms) from Eq. (1.8). In other words, the crossbar switch scheduling problem is considered as N-rook problem where a rook commands horizontally and vertically. The positions of the rooks are equivalent to the state of the traffic matrix. The goal of the N x N crossbar switch scheduling problem is to place N mutually nonattacking rooks on an N x N chessboard. The maximum number of rooks is given by N^2.

The simulator has been developed on a Macintosh SE/30 and a DEC3100 workstation. Remember that the simulator can be easily implemented by Program1, Program2, or Program3 as shown in Chapter 1. The demonstrated simulator here is completely based on Program2 in Fig. 1-7 except the input/output function where the hysteresis McCulloch-Pitts neuron model is used. The simulator is to simulate the synchronous parallel neural network model. The motion equation in Eq. (2.1) is inserted in Program2. Fig. 2-2 shows the relationship between the average number of iteration steps, the problem size, and the band size of hysteresis. The hysteresis band size is given by the hysteresis band size=|UTP|=|LTP|. When no hysteresis is given to each neuron, it usually takes more than 5000 iteration steps or does not converge to the global minimum at all.

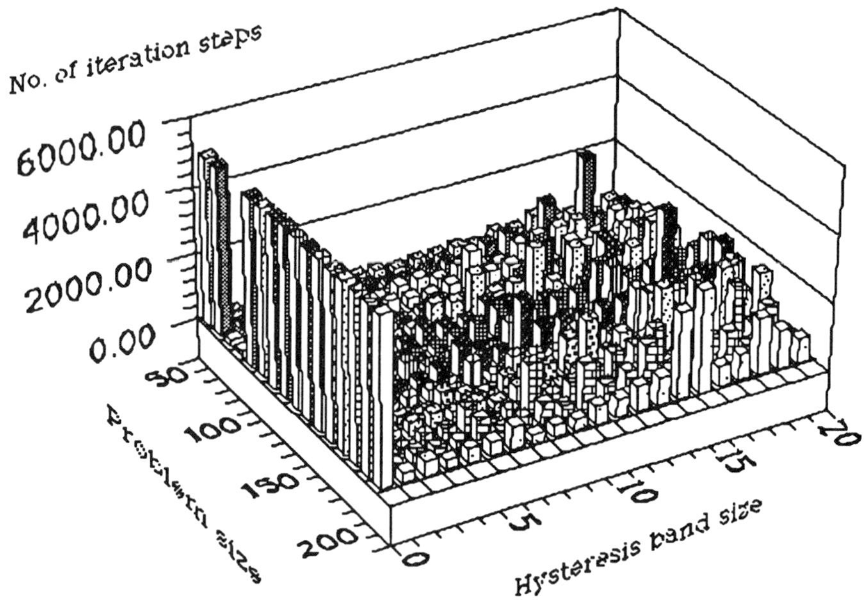

Fig. 2-2 (a) The relationship between the number of iteration steps, the problem size, and the hysteresis band size.

Fig. 2-2 (b) zooms in the part of Fig. 2-2a where a range of the hysteresis band size is one through five. Fig. 2-2a and 2-2b show that the state of the system can converge to the global minimum in nearly O(1) time. The hysteresis band size=1, 2, or 3 gives the relatively good convergence to the global minimum. Fig. 2-3 shows the relation between the average number of iteration steps and the problem size where 10 x 10 crossbar problems through 150 x 150 crossbar problems were investigated. Note that 100 simulation runs were performed to obtain the one average number of iteration steps per problem. The simulation result depicts that the state of the system converged to the acceptable solution within several hundred iteration steps. The solution quality of a hysteresis McCulloch-Pitts binary neural network with small hysteresis does not degrade with the problem size as far as we have observed.

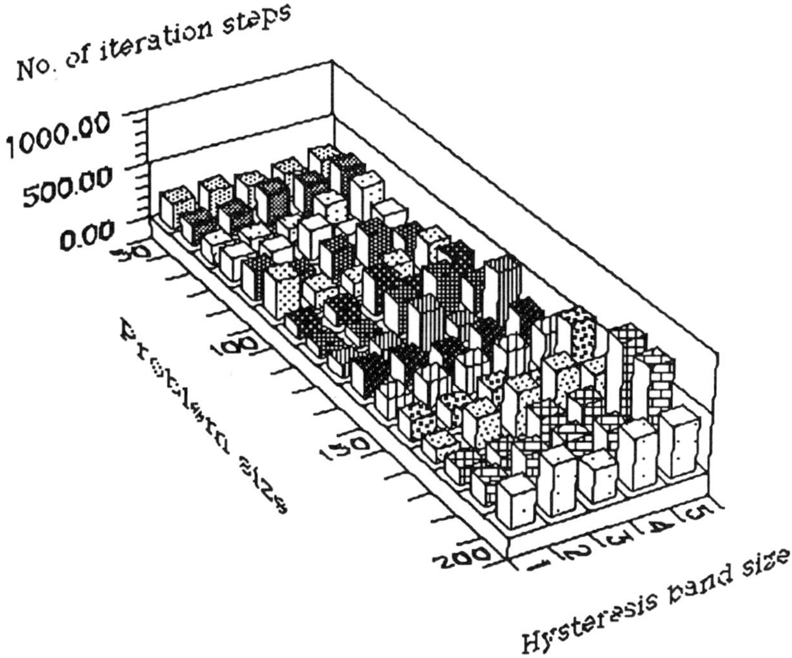

Fig. 2-2 (b) The relationship between the number of iteration steps, the problem size, and the hysteresis band size.

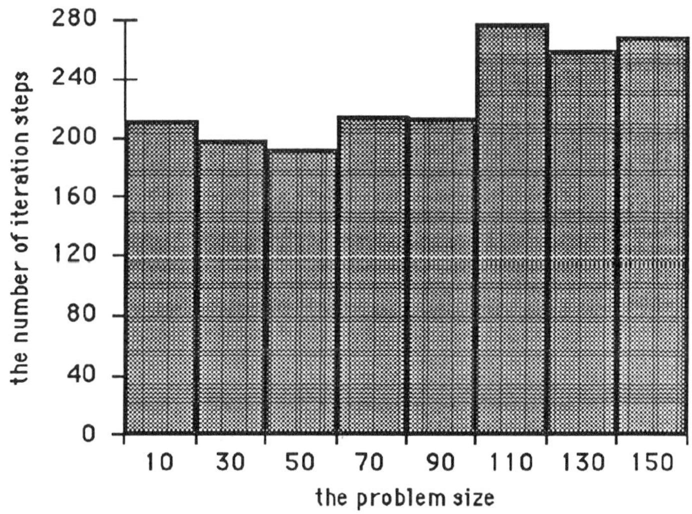

Fig. 2-3 The relationship between the average number of iteration steps and the problem size when the hysteresis band size is 3.

Fig. 2-4a, and 2-4b show the 64 x 64 crossbar problem and one of the solutions respectively. The following parameters were used in our simulation: A=1, B=1 if (t mod 10)<2, 0 otherwise, U_min=-800, and U_max=200. Based on our simulation results the hysteresis in individual neurons effectively suppresses the oscillatory behaviors of neural dynamics. Consequently the convergence time to the solution is shortened.

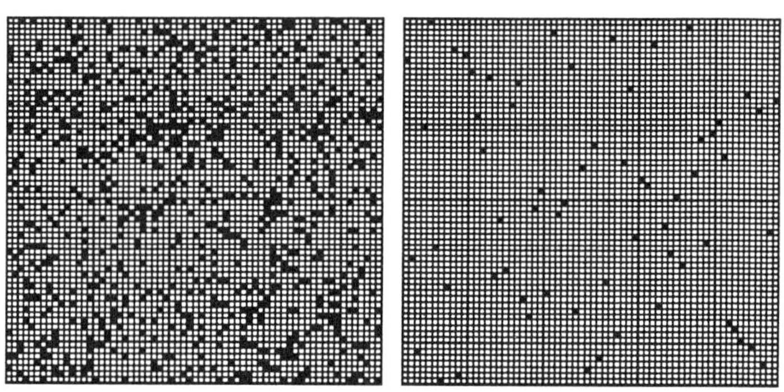

(a) The problem (b) one of the solutions

Fig. 2-4 The 64 x 64 crossbar switch problem and one of the solutions

2.3 REFERENCES

Chen W., Mavor J., Denyer P. B., Renshaw D. (1990) Traffic routing algorithm for serial superchip system customisation. IEE Proc., 137, Pt. E, 1

Hoffman G. W., Benson M. W. (1986) Neurons with hysteresis form a network that can learn without any changes in synaptic connection strengths. Proc. of AIP Conf. on Neural Networks for Computing (Ed. J. S. Denker, AIP 1986)

Hopfield J. J. (1982) Neural networks and physical systems with emergent collective computational abilities. Proc. Natl. Acad. Sci. USA, 79, 2554-2558

Hopfield J. J., Tank D. (1986) Computing with neural circuits: A model. Science, 233, 625-633

Inukai T. (1979) An efficient SS/TDMA time slot assignment algorithm. IEEE Trans. on Commun., 27, 1449-1455

Keeler J. D., Pichler E. E., Ross J. (1989) Noise in neural networks: thresholds, hysteresis, and neuromodulation of signal-to-noise. Proc. Natl. Acad. Sci. USA, 86, 1712-1716

Marrakchi A., Troudet T. (1989) A Neural Net Arbitrator for Large Crossbar Packet-Switches. IEEE Trans. on Circuits and Systems, 36, 7

McCulloch W. S., Pitts W. (1943) A logical calculus of the ideas imminent in nervous activity. Bull. Math. Biophys., 5, 115-133

Rose C. (1989) Rapid Optimal Scheduling for Time-Multiplex Switches Using a Cellular Automaton. IEEE Trans. on Commun., 37, 500-509

Segundo J. P., Martinez O. D. (1985) Dynamic and Static hysteresis in Crayfish stretch receptors. Biological Cybernetics, 52, 291-296

Simic B. G. (1986) SPIE, 634, 195

Takefuji Y., Lee K. C. (1989) A near-optimum parallel planarization algorithm. Science, 245, 1221-1223

Takefuji Y., Lee K. C. (1990a) A parallel algorithm for tiling problems. IEEE Trans. on Neural Networks, 1, 1, 143-145

Takefuji Y., Lee K. C. (1990b) A super parallel sorting algorithm based on neural networks. IEEE Trans. on Circuits and Systems, 37, 11

Takefuji Y., Lee K. C. (1991) An hysteresis binary neuron: a model suppressing the oscillatory behaviors of neural dynamics. Biological Cybernetics, 64, 353-356

2.4 EXERCISES

1. Generate several traffic matrices with different densities and observe the behavior of the system by changing the hysteresis band size with or without the hill-climbing term.
2. Compare the solution quality and the computation time of the proposed algorithm with that of the best known algorithm.
3. Build a simulator to simulate the synchronous parallel N x N crossbar model as if P x Q processing elements are used on a parallel machine for P, Q<N.
4. Discuss why hysteresis in the neuron model is able to suppress the oscillatory behavior in neural dynamics.
5. Survey other crossbar switch problems and solve one of their problems.

Chapter 3

FOUR-COLORING AND K-COLORABILITY PROBLEMS

The computational energy is presented for solving a four-coloring map problem. The map-coloring problem is defined that one wants to color the regions of a map in such a way that no two adjacent regions (that is, regions sharing some common boundary) are of the same color. This Chapter presents a neural network parallel algorithm based on the McCulloch-Pitts binary neuron model. A 4 x n neural array is used to color a map of n regions where each neuron as a processing element performs the proposed motion equation. The capability of the proposed system is demonstrated through a large number of simulation runs. The parallel algorithm is modified for solving the k-colorability problem. This Chapter is based on our paper published in IEEE Transactions on Circuits and Systems (Takefuji and Lee 1991)

3.1 INTRODUCTION

A map maker colors adjacent countries with different colors so that they may be easily distinguished. This is not a problem as long as one has a large number of colors. However it is more difficult with a constraint that one must use the minimum number of colors required for a given map. It is still easy to color a map with a small number of regions. In the early 1850's Francis Guthrie was interested in this problem who brought it to the attention of Augustus De Morgan. Since then many mathematicians including Arthur Kempe, Peter Tait, Percy Heawood, and others tried to prove the problem that any planar simple graph can be colored with four colors. A four-color problem is defined that one wants to color the regions of a map in such a way that no two adjacent regions (that is, regions sharing some common boundary) are of the same color. In August 1976, Kenneth Appel and Wolfgang Haken presented their work to members of the American Mathematical Society (Appel and Haken 1977). They showed a computer-aided proof of the four-color problem. However their coloring was based on the sequential method so that it took many hours to solve a large problem. Their computation time may be proportional to $O(n^2)$ where n is the number of regions.

Few parallel algorithms have been reported. E. D. Dahl (Dahl 1987), Moopenn et al. (Moopenn et al. 1987), and Thakoor et al. have presented the first neural network for map K-colorability problems (Thakoor et al. 1987). In our algorithm the four-color problem is solved by a 4 x n two dimensional neural array where n is the number of regions to be colored.

3.2 FOUR-COLORING NEURAL NETWORK

Four colors can be simply expressed by four neurons. For example red, yellow, blue, and green are represented by 1000, 0100, 0010, and 0001 respectively. A single region requires four neurons for the single-color assignment. Therefore a 4 x n two dimensional neural array is needed for coloring n regions of a map. For example, nine regions are colored by four colors as shown in Fig. 3-1. The neural representation for the nine-region map problem is given in Fig. 3-2 where a 4 x 9 neural array is used. Black squares indicate fired neurons and while squares for non-fired neurons. The solution in Fig. 3-1 corresponds to the state of neural array in Fig. 3-2. Fig. 3-3 shows the adjacency matrix of the nine-region map which gives the boundary information between regions. Black squares depict the adjacency between regions.

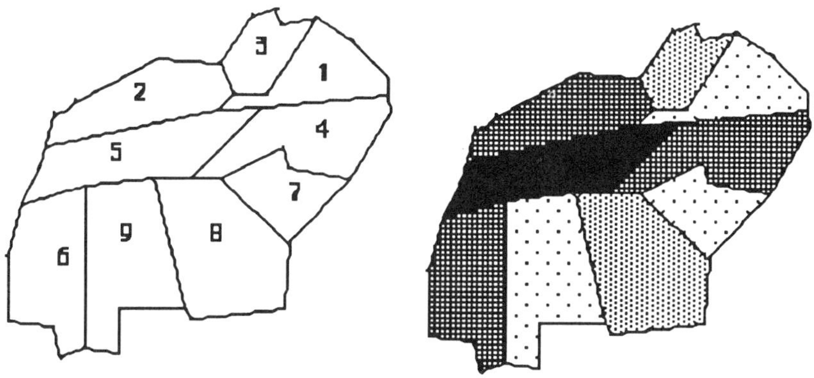

Fig. 3-1 Nine-region map and 4-coloring

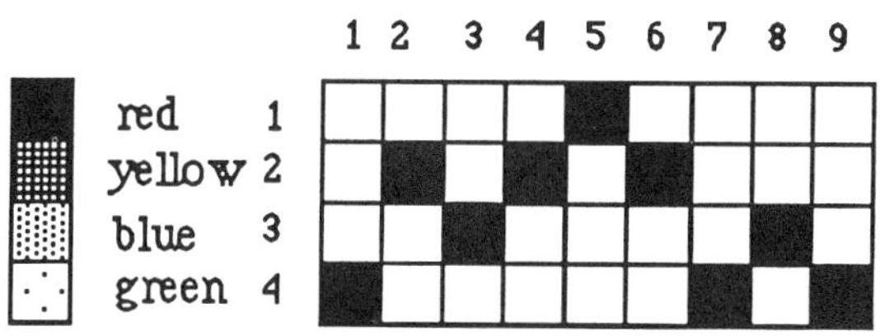

Fig. 3-2 Neural representation for 4-coloring map in Fig. 3-1

Fig. 3-3 Adjacency matrix of the map in Fig. 3-1

In order to consider in such a way that no two adjacent regions are of the same color, the basic energy function is given by:

$$E = \frac{A}{2}\sum_{X=1}^{n}\left(\sum_{i=1}^{4} V_{Xi}-1\right)^2 + B\sum_{X=1}^{n}\sum_{\substack{Y=1\\Y\neq X}}^{n}\sum_{i=1}^{4} d_{XY}V_{Xi}V_{Yi} \tag{3.1}$$

where d_{XY} is 1 if regions X and Y are adjacent to each other, 0 otherwise. Note that A and B are constant numbers. The adjacency matrix is an n x n array where d_{XY} is a single element. Note that V_{Xi} is the output of the *i*th neuron in the *X*th region. The first term corresponds to the column constraint in the neural array which forces one and only one neuron's output to be nonzero. The last term describes the boundary violation between regions. If X and Y regions have a common boundary (d_{XY}=1), then X and Y regions should not have the same color i.

The motion equation of the *i*th color of the *X*th region is given by:

$$\frac{dU_{Xi}}{dt} = -\frac{\partial E}{\partial V_{Xi}} \tag{3.2}$$

Eq. (3.1) and (3.2) generate the equation of the *i*th color of the *X*th region by:

$$\frac{dU_{Xi}}{dt} = -A\left(\sum_{j=1}^{4} V_{Xj}-1\right) - B\sum_{\substack{Y=1\\Y\neq X}}^{n} d_{XY}V_{Yi} \tag{3.3}$$

We have introduced the hill-climbing term which allows the state of the system to escape from the local minimum and to converge to the global minimum. The hill-climbing term and the second term are normalized. The final motion equation of the *i*th color of the *X*th region is given by:

$$\frac{dU_{Xi}}{dt} = -A\left(\sum_{j=1}^{4} V_{Xj} - 1\right) - B\sum_{\substack{Y=1\\Y\neq X}}^{n} d_{XY}V_{Yi}\sum_{K=1}^{n} d_{YK}$$

$$+C\cdot h\left(\sum_{j=1}^{4} V_{Xj}\right)\left(C_1\sum_{K=1}^{n} d_{XK} + C_2\frac{\sum_{K=1}^{n}\sum_{Y=1}^{n} d_{XY}d_{YK}}{\sum_{K=1}^{n} d_{XK}}\right) \qquad (3.4)$$

The coefficients of the second term are normalized. Note that C, C_1, and C_2 are constant parameters. The last term is the hill-climbing term where h(x) is 1 if x=0, 0 otherwise. It performs the excitatory force only when all of V_{Xj} are zeros. The hill-climbing term allows the state of the system to escape from the local minimum and increases the frequency to converge to the global minimum.

The simulator was developed based on Program2 in Fig. 1-8 where the following coefficients are used: $A=B=C=C_1=C_2=1$. Fig. 3-4 shows the US-continental map which consists of 48 states where the adjacency matrix is given by a 48 x 48 array.

Fig. 3-5a and 3-5b show the convergence of the 48-state US-map neural network to a four-color solution respectively. The linear dimension of each rectangle is proportional to the value of ΔU_{Xi}, U_{Xi}, or V_{Xi}. Black and white rectangles indicate positive and negative respectively. Fig. 3-5a shows the intermediate state of 192 neurons after the first iteration.

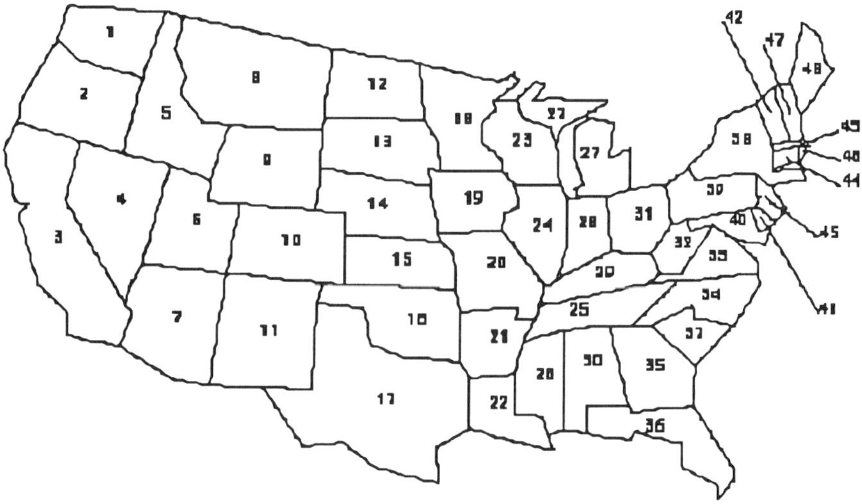

Fig. 3-4 US-continental map

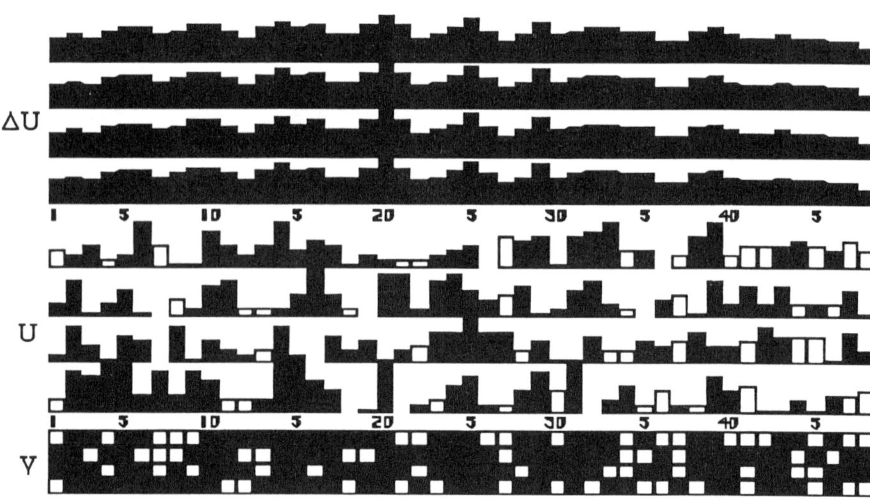

Fig. 3-5 (a) The convergence of the 48-state US map neural network to a four-coloring solution.

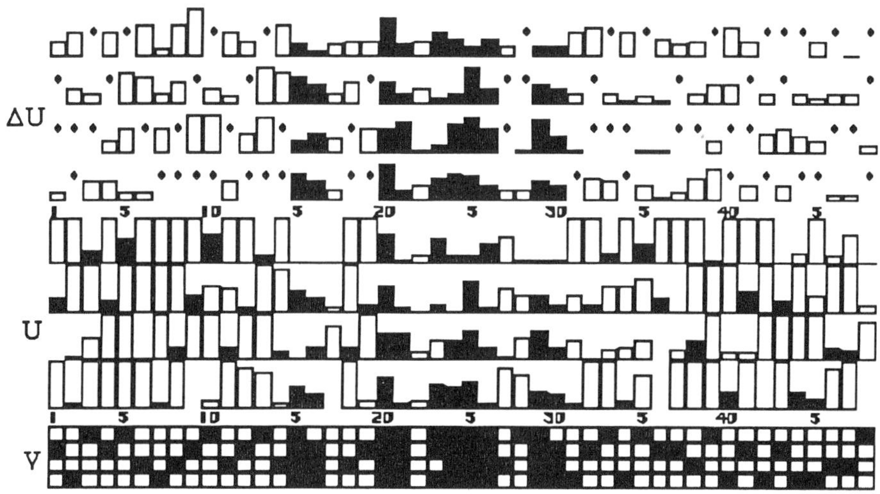

Fig. 3-5 (b) The convergence of the 48-state US map neural network to a four-coloring solution. This shows the intermediate state of 192 neurons after the tenth iteration.

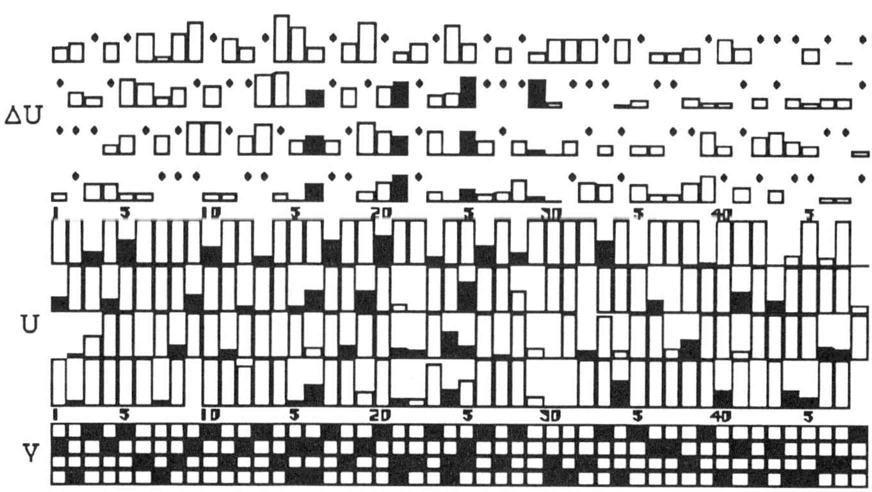

Fig. 3-5 (c) The convergence of the 48-state US map neural network to a four-coloring solution. This shows the intermediate state of 192 neurons after the twentieth iteration.

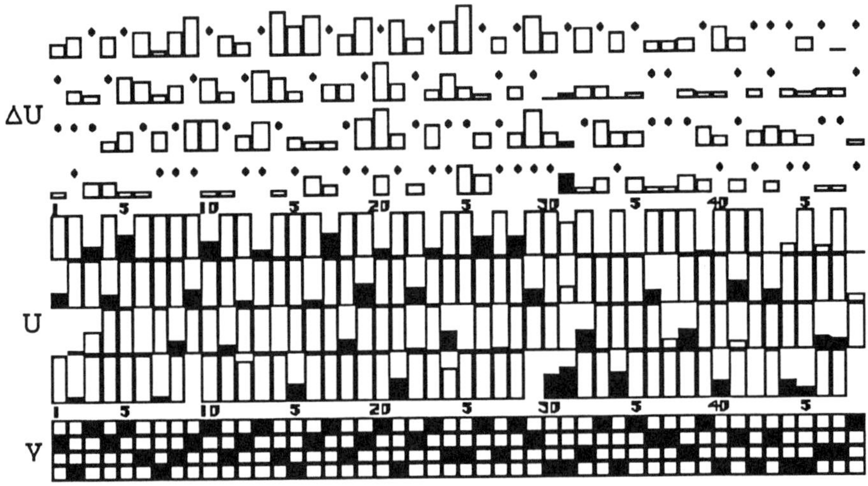

Fig. 3-5 (d) The convergence of the 48-state US map neural network to a four-coloring solution. This shows the final state of 192 neurons after the thirty fifth iteration.

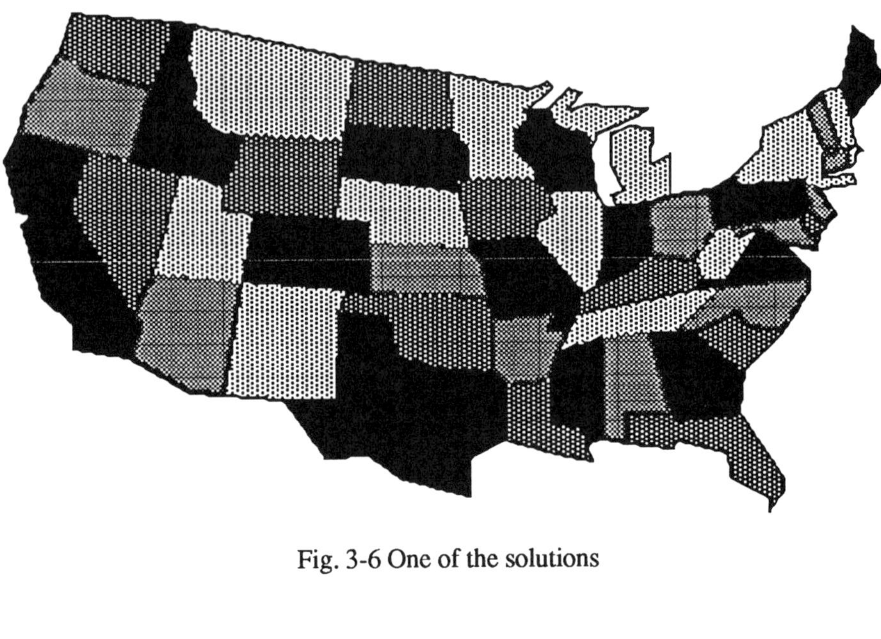

Fig. 3-6 One of the solutions

Fig. 3-6 depicts the solution which is decoded from the final state of 192 neurons in Fig. 3-5d. Fig. 3-7 describes one of the solutions for a 210-country map four-coloring problem where the problem is taken from the example of Appel and Haken's experiments. The state of the system always converged to the global minimum as far as we have observed through more than one thousand simulation runs. Fig. 3-8 shows the relationship between the frequency and the number of iteration steps using several hundred simulation runs in the 210-country map problem. The average number of iteration steps is 820.

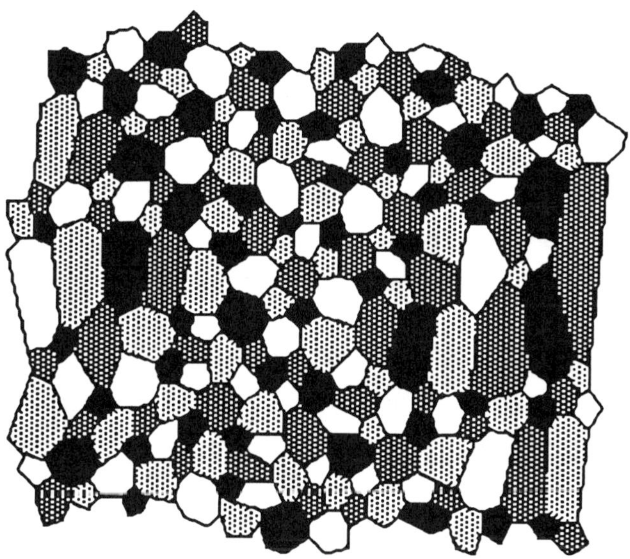

Fig. 3-7 One of the solutions for the 210-country map problem

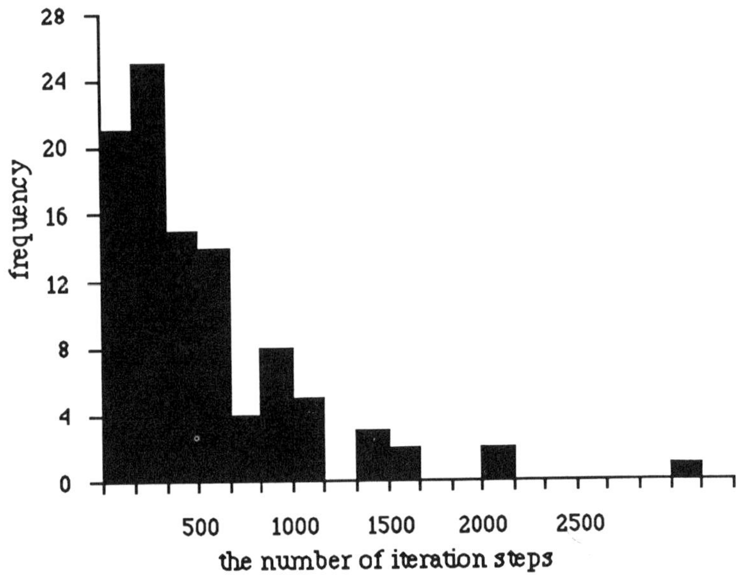

Fig. 3-8 Relationship between the frequency and the number of iteration steps in the 210-country map coloring problem

Fig. 3-9 shows one of the solutions for a 430-country map problem. Fig. 3-10 depicts the frequency versus the number of iteration steps using several hundred simulation runs for the 430-country map problem. The average number of iteration steps is 1000. Fig. 3-8 and Fig. 3-10 indicate that the problem size does not strongly influence the number of iteration steps to converge to the global minimum.

Through more than one thousand sets of simulation runs we have observed that the problem size does not strongly influence the number of iteration steps to converge to the global minimum. The proposed scaling method for the coefficients in the motion equation is quite effective which allows the state of the system to escape from the local minimum and to converge to the global minimum.

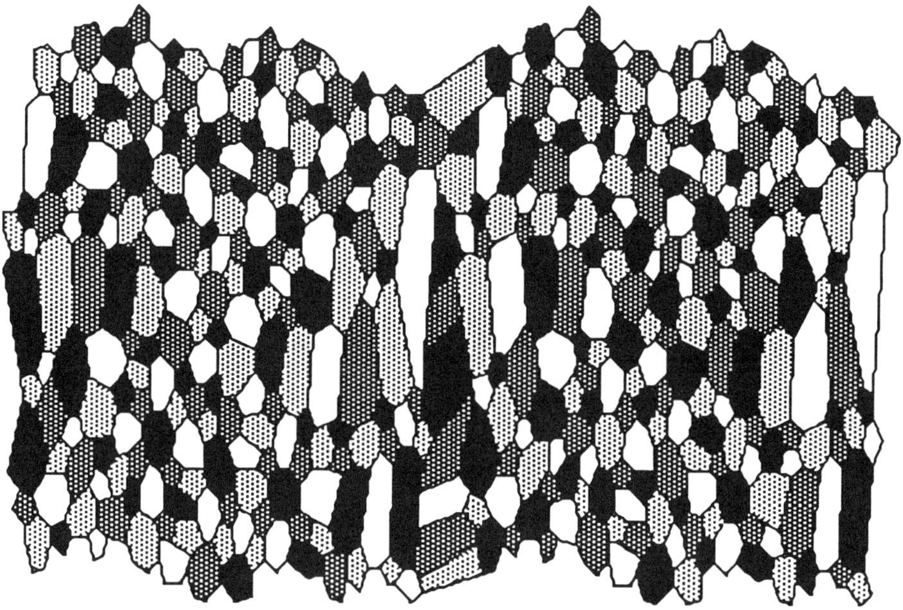

Fig. 3-9 One of the solutions for the 430-country map problem

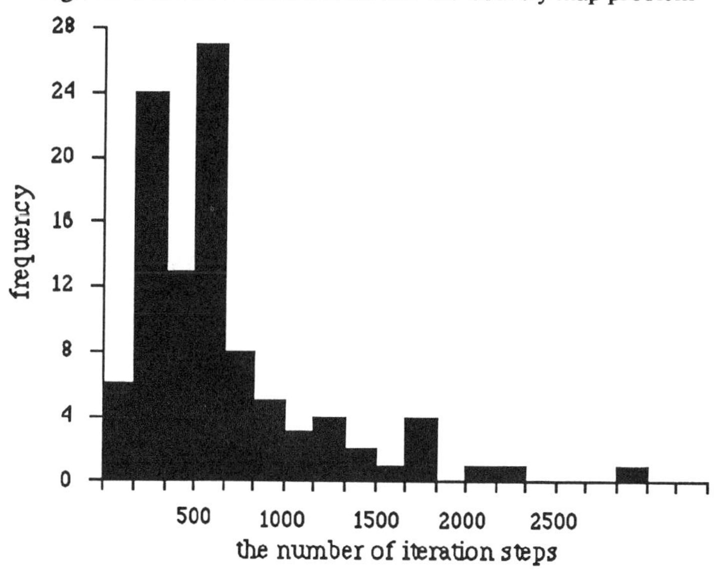

Fig. 3-10 Relationship between the frequency and the number of iteration steps in the 430-country map coloring problem

3.3 K-COLORABILITY NEURAL NETWORK

The motion equation in Eq. (3.4) is modified for the three-color map problem. For the three-color map problem a 3 x n neural array is prepared. It is given by:

$$\frac{dU_{Xi}}{dt} = -A\left(\sum_{j=1}^{3} V_{Xj}-1\right) - B\sum_{\substack{Y=1\\Y\neq X}}^{n} d_{XY}V_{Yi}\sum_{K=1}^{n} d_{YK}$$

$$+C\cdot h\left(\sum_{j=1}^{3} V_{Xj}\right)\left(C_1\sum_{K=1}^{n} d_{XK}+C_2\frac{\sum_{K=1}^{n}\sum_{Y=1}^{n} d_{XY}d_{YK}}{\sum_{K=1}^{n} d_{XK}}\right) \tag{3.5}$$

Fig. 3-11 shows three-colorability for the 48-state US-map problem. Our simulator always indicates that two conflicts exist in a map: a conflict around California and the other conflict around West Virginia.

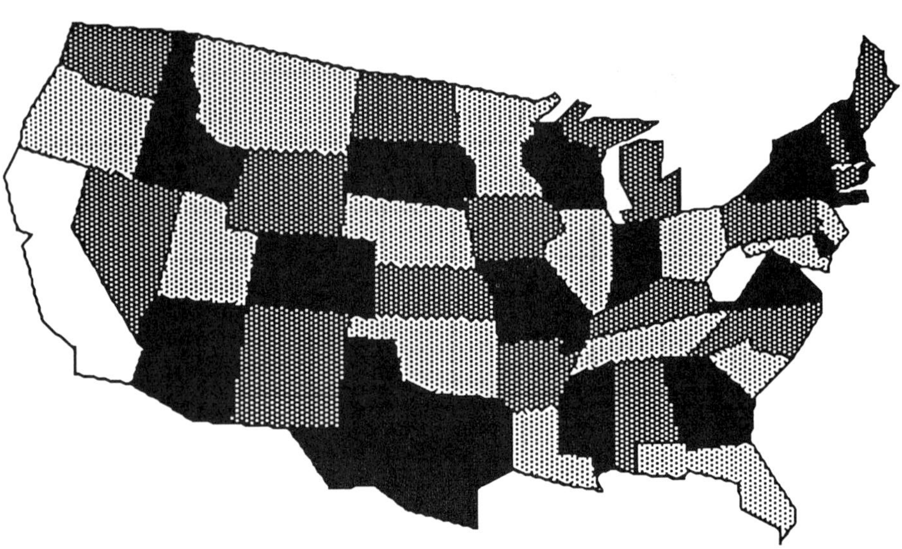

Fig. 3-11 3-colorability of the 48-state US map problem

In general an m x n neural array is required for the m-color problem where n is the number of regions. K-colorability problems belong to the NP-complete problem (Garey and Johnson 1979). The motion equation of the *i*th color of the *X*th region for the m-color-n-region problem is given by:

$$\frac{dU_{Xi}}{dt} = -A\left(\sum_{j=1}^{m} V_{Xj} - 1\right) - B\sum_{\substack{Y=1\\Y\neq X}}^{n} d_{XY}V_{Yi}\sum_{K=1}^{n} d_{YK}$$

$$+C\cdot h\left(\sum_{j=1}^{m} V_{Xj}\right)\left(C_1\sum_{K=1}^{n} d_{XK} + C_2\frac{\sum_{K=1}^{n}\sum_{Y=1}^{n} d_{XY}d_{YK}}{\sum_{K=1}^{n} d_{XK}}\right) \quad (3.6)$$

We introduced the computational energy function for the four-color map problem and the k-color map problem. With the hill-climbing term and the normalization, the proposed motion equation forces the state of the system to escape from the local minimum and to converge to the global minimum. A large number of simulation runs were performed to verify our algorithm. The simulation result supports that the problem size does not influence the number of iteration steps to converge to the global minimum.

3.4 REFERENCES

Appel K., and Haken W., (Oct. 1977), The solution of the four-color-map problem, Scientific American, pp.108-121.

Dahl E. D., (1987), Neural Network Algorithm for an NP-Complete Problem: Map and Graph Coloring, Proc. of the first Int. Conf. on Neural Networks, vol. III, pp.113-120.

Garey M. R., and Johnson D. S., (1979), Computers and Intractability: A Guide to the theory of NP-completeness, (W. H. Freeman and Company).

Moopenn A. et al., (1987), A neurocomputer based on an analog-digital hybrid architecture, Proc. of the first Int. conf. on neural networks, vol. III, pp.479-486.

Thakoor A. P. et al., (1987), Electronic hardware implementations of neural networks, Applied Optics, vol. 26, pp.5085-5092.

Takefuji Y., and Lee K. C., (1991), Artificial neural networks for four-coloring problems and k-colorability problems, IEEE Trans. on Circuits and Systems, 38, 3, pp.326-333.

3.5 EXERCISES

1. Build two simulators based on Eq. (3.3) and E. (3.4) and observe the effect of the hill-climbing and that of the normalization in Eq. (3.4)
2. Explain a role of the normalization in Eq. (3.4) and how it is obtained.
3. Write a four-coloring game which not only generates a local minimum and contains the conflict(s) but also provides a solution from the local minimum by changing the minimum number of colored regions. The goal of the game is to allow a player to guess how many minimum colored regions must be updated to satisfy the four-coloring problem.
4. Discuss the relationship between the cellular phone frequency assignment problem and the k-colorability problem and find other applications based on coloring problems.
5. Write a simulator discussed above.

Chapter 4

GRAPH PLANARIZATION PROBLEMS

In this Chapter a near-optimum parallel planarization algorithm is introduced. The proposed system composed of an N x N neural network array not only generates a near maximal planar subgraph from a nonplanar graph or a planar graph but also embeds the subgraph on a single plane within O(1) time. The algorithm can be used in designing printed circuit boards and routing very large scale integrated (VLSI) circuits. This Chapter is based on a paper published in Science (Takefuji and Lee 1989).

4.1 INTRODUCTION

Maximal planarization of a planar or nonplanar graph is an important problem in designing printed circuit boards and routing VLSI circuits. A graph is planar if it can be drawn on a single plane with no two edges crossing each other except at their end vertices. If a given circuit is planar, it can be wired on a single layer. If a given circuit is nonplanar, we would like to maximize the number of edges to be planarized and minimize the number of edges to be removed from a nonplanar graph. To yield a maximal planar subgraph from a nonplanar graph is an NP-complete problem (Garey and Johnson 1979).

Two tasks must be accomplished to solving the graph planarization problem.

One is to verify whether the circuit is planar or not and then to extract a planar subgraph from the nonplanar graph in order to embed it on a plane, called the planarity testing. In 1989 Jayakumar, Thulasiraman, and M. N. S. Swamy have proposed the $O(N^2)$ near-optimum planarity testing algorithm (Jayakumar et al. 1989) where N is the number of vertices. The other task is to embed the planar subgraph on a plane, called plane embedding where we want to maximize the number of edges in the graph. The plane embedding algorithm was reported by R. E. Tarjan in 1971 (Tarjan 1971). Unfortunately few parallel algorithms have been reported to the graph planarization problem.

This Chapter introduces the new parallel graph planarization algorithm. The algorithm not only yields a maximal planar subgraph from a nonplanar graph or a planar graph but also embeds the subgraph on a single layer within O(1) time. In other words, the algorithm provides the necessary and sufficient routing information for embedding a planar subgraph from the given nonplanar or planar graph. The algorithm uses 2M processing elements (neurons) for the M-edge graph planarization problem where a McCulloch-Pitts binary neuron model is used.

4.2 NEURAL REPRESENTATION AND MOTION EQUATIONS

It is very difficult to describe the topological information, for example, two edges crossing each other. Consider the simple undirected graph composed of 4 vertices and 6 edges as shown in Fig. 4-1a.

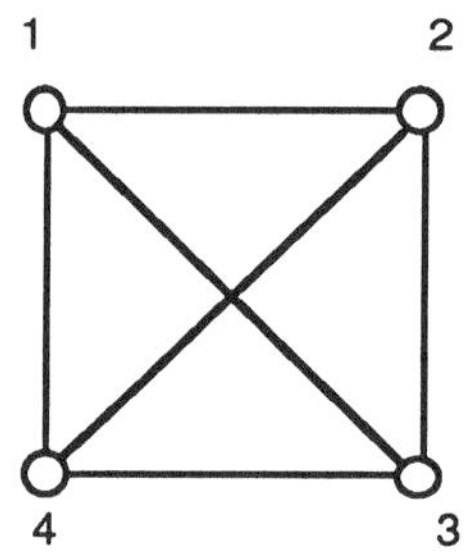

Fig. 4-1 (a) 4-vertex 6-edge graph

The graph is planar as long as two edges (1,3) and (2,4) do not cross each other. Fig. 4-1b and 4-1c show two possible planar graphs. The key question is how to express such topological information to be able to represent the two-edge-crossing violation and to embed those graphs.

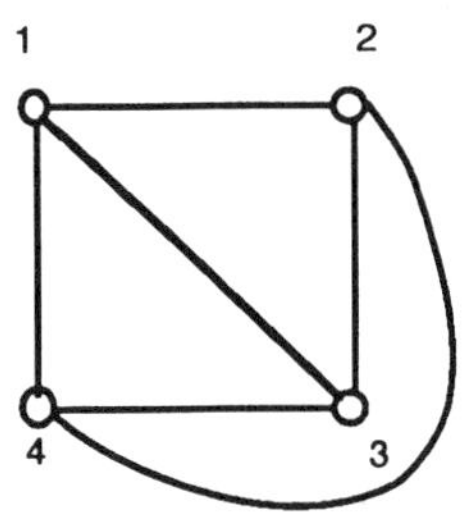

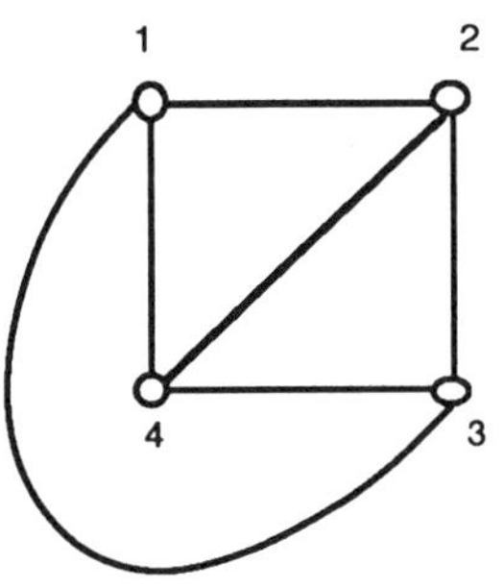

Fig. 4-1 (b) 4-vertex 6-edge graph   Fig. 4-1 (c) 4-vertex 6-edge graph

In this Chapter the single-row routing representation is used. In the single-row routing representation, connection is established by either a upper edge or a lower edge. The difficult topological information is expressed by the single-row routing representation. Two neurons (Vup_{ij} and $Vdown_{ij}$) are provided to express a upper line connection or a lower line connection between the *i*th vertex and the *j*th vertex. For example, the state of two neurons (Vup_{ij}=1 and $Vdown_{ij}$=0) indicates that the edge (i,j) is established by embedding the upper line connection. The following states (Vup_{ij}=$Vdown_{ij}$=0), (Vup_{ij}=0 and $Vdown_{ij}$=1), and (Vup_{ij}=$Vdown_{ij}$=1) imply no connection, lower line connection, and double-line-connection violation respectively. Fig. 4-1d shows one of possible planar graphs based on the single-row routing representation.

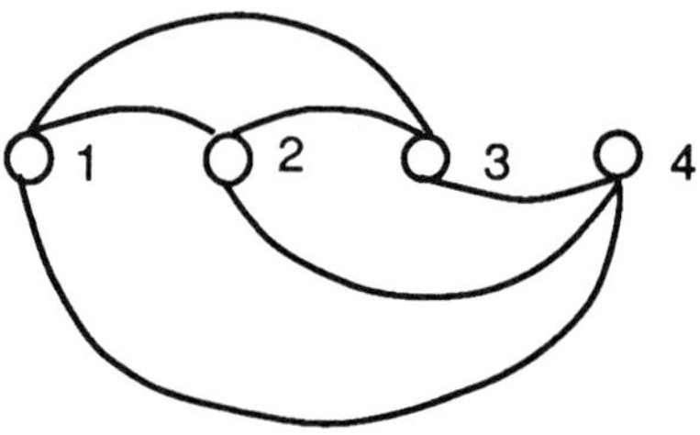

Fig. 4-1 (d) graph in single-row representation

The number of edges in a given graph determines the number of required neurons. The system requires 2M neurons where M is the number of edges in a graph. For example, the graphs shown in Fig. 4-1d can be represented by a 2 x 6 two-dimensional neural network array as shown in Fig. 4-2. Fig. 4-1e shows the possible edge intersections where only two edges (1,3) and (2,4) intersect each other. The number of required processing elements for planarization of a graph with M edges is actually much less than 2M. In Fig. 4-1e four neurons are needed because two edges intersect each other.

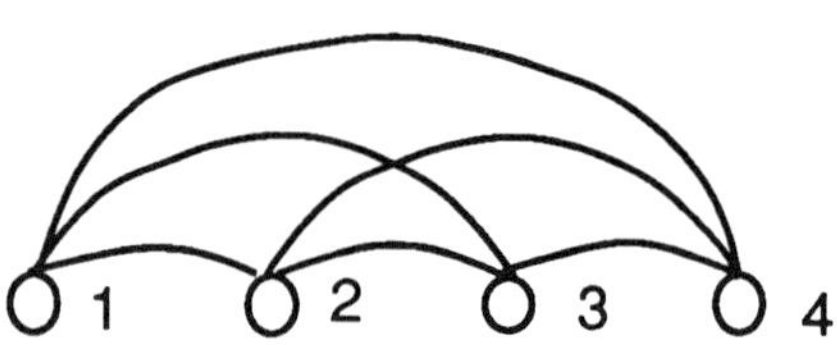

Fig. 4-1 (e) Graph in single-row representation

Vup_{12}	Vup_{13}	Vup_{14}	Vup_{23}	Vup_{24}	Vup_{34}
$Vdown_{12}$	$Vdown_{13}$	$Vdown_{14}$	$Vdown_{23}$	$Vdown_{24}$	$Vdown_{34}$

1	1		1		
		1		1	1

Fig. 4-2 Single-row representation of the graph in Fig. 4-1d

There are two kinds of forces performing in the artificial neural network; excitatory and inhibitory forces. In the graph planarization problem the excitatory force means that if an edge (i,j) exists in a given graph then the *ij*th neuron to represent embedding the connection line in a plane is encouraged to fire. The inhibitory force means that the neurons which violate the two-edge-crossing condition are discouraged to fire. The two-edge-crossing violation condition is expressed by the following:

If $Vup_{lm}=1$ and ($l<i<m<j$ or $i<l<j<m$) then the *ij*th up-neuron should not be fired. In other words, Vup_{ij} should not be 1. If $Vdown_{lm}=1$ and ($l<i<m<j$ or $i<l<j<m$) then $Vdown_{ij}$ should not be 1. Fig. 4-3 describes the edge-violation conditions for Vup_{ij} and $Vdown_{ij}$. The necessary and sufficient conditions of our violation conditions are given by the concise functions:

$\sum_{l} \sum_{\substack{m \\ l<m}} f(l,i,m)f(i,m,j)Vup_{lm}$ for Vup_{ij} neuron is shown in Fig. 4-3a where the function $f(L,M,R)$ is 1 if $L<M<R$, 0 otherwise.

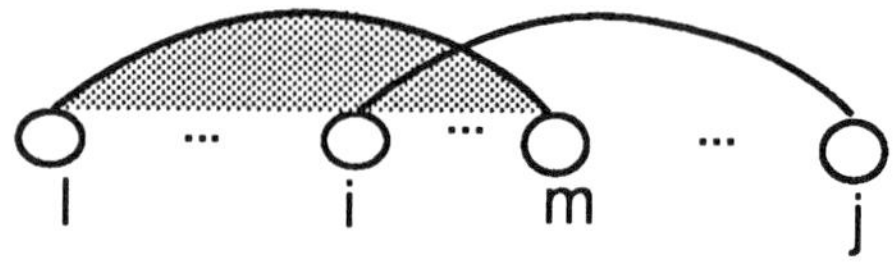

Fig. 4-3 (a) One of violation conditions in single-row representation

$\sum_{l} \sum_{\substack{m \\ l<m}} f(i,l,j)f(l,j,m)\,Vup_{lm}$ for Vup_{ij} neuron is shown in Fig. 4-3b.

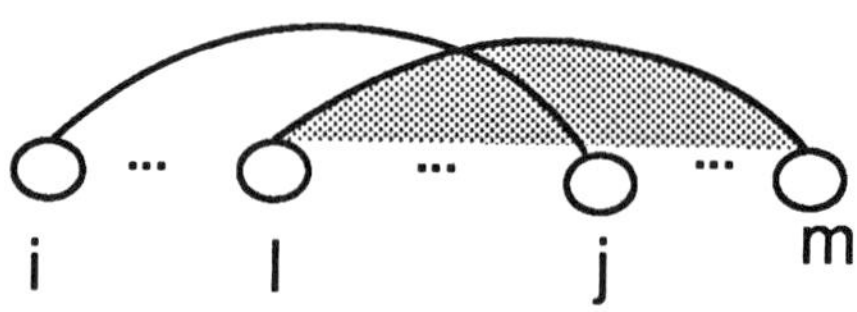

Fig. 4-3 (b) One of violation conditions in single-row representation

$\sum_{l} \sum_{\substack{m \\ l<m}} f(l,i,m)f(i,m,j)\,Vdown_{lm}$ for $Vdown_{ij}$ neuron is given in Fig. 4-3c.

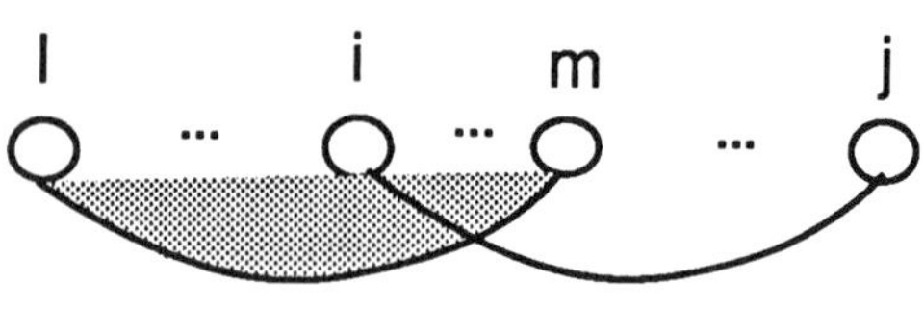

Fig. 4-3 (c) One of violation conditions in single-row representation

$\sum_{l} \sum_{\substack{m \\ l<m}} f(i,l,j)f(l,j,m)\,Vdown_{lm}$ for $Vdown_{ij}$ neuron is given in Fig. 4-3d.

Fig. 4-3 (d) One of violation conditions in single-row representation

Therefore the motion equation of the *ij*th up-neuron is given by:

$$\frac{dUup_{ij}}{dt} = -A(Vup_{ij}+Vdown_{ij} - 1) -B\sum_{l}\sum_{\substack{m\\l<m}} f(l,i,m)f(i,m,j)Vup_{lm}$$
$$-B\sum_{l}\sum_{\substack{m\\l<m}} f(i,l,j)f(l,j,m)Vup_{lm} \qquad (4.1)$$

The first term in Eq. (4.1) performs the excitatory force as long as no two-neuron is fired (Vup_{ij}=$Vdown_{ij}$=0). If (Vup_{ij}=1&$Vdown_{ij}$=0) or (Vup_{ij}=0&$Vdown_{ij}$=1) then the first term will take no action. If (Vup_{ij}=$Vdown_{ij}$=1) then it will perform the inhibitory force. The second and the last term in Eq. (4.1) are always the inhibitory forces.

The motion equation of the *ij*th down-neuron is given by:

$$\frac{dUdown_{ij}}{dt} = -A(Vup_{ij}+Vdown_{ij} - 1) -B\sum_{l}\sum_{\substack{m\\l<m}} f(l,i,m)f(i,m,j)Vdown_{lm}$$
$$-B\sum_{l}\sum_{\substack{m\\l<m}} f(i,l,j)f(l,j,m)Vdown_{lm} \qquad (4.2)$$

In general the planarization problem for a graph with N vertices and M edges can be solved by an N x N neural network. The motion equation of the *ij*th X-neuron where X is either up or down is given by:

$$\frac{dUx_{ij}}{dt} = -A(Vup_{ij}+Vdown_{ij} - C_{ij}) -B\sum_{l}\sum_{\substack{m\\l<m}} f(l,i,m)f(i,m,j)Vx_{lm}$$
$$-B\sum_{l}\sum_{\substack{m\\l<m}} f(i,l,j)f(l,j,m)Vx_{lm} \qquad (4.3)$$

where C_{ij} is 1 if the edge (i,j) exists in the given graph, 0 otherwise.

Consider the nonplanar graph given by Jayakumar and et. al (Jayakumar et al. 1989) in Fig. 4-4. The graph has 10 vertices and 22 edges. The latest experiment by Jayakumar and et. al. showed that 19 edges were chosen by their $O(N^2)$ algorithm to construct the maximal planar subgraph as shown in Fig. 4-5. Remember that their algorithm does not embed the generated subgraph on a plane but chooses edges for construction of the maximal planar subgraph.

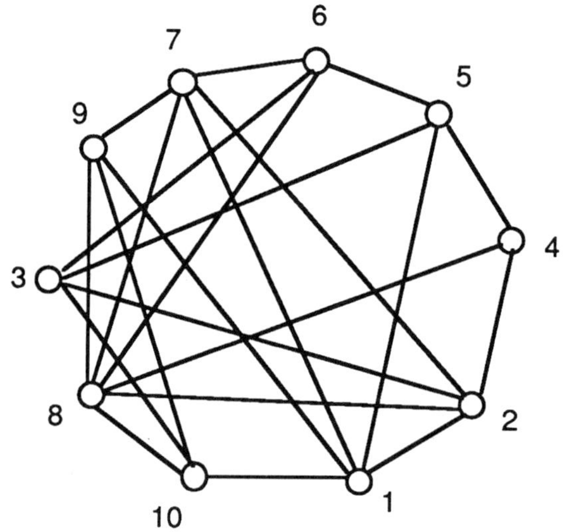

Fig. 4-4 Nonplanar graph with 10 vertices and 22 edges

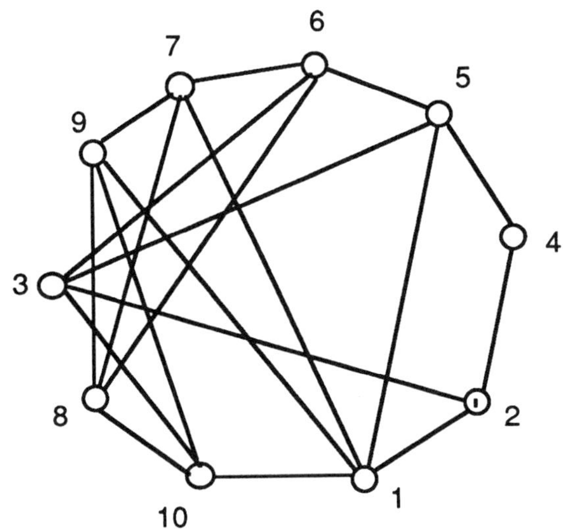

Fig. 4-5 Maximal planar subgraph by Jayakumar

We have developed the simulator based on Program2 as discussed in Chapter 1.4 in Fig. 1-8. First we solved the same nonplanar graph problem as shown in Fig. 4-4. When the coefficients A=2 and B=1, the unit time $\Delta t=10^{-5}$ were used for Eq. (4-1) and (4-2), and the initial values of $Uup_{ij}(t)$ and $Udown_{ij}(t)$ where i=1,...,N and j=1,...,N were randomized in the range of -1/10000 to 0, the state of our system converged to the global minimum in the 14th iteration. Fig. 4-6a and 4-6b describe the state of the system at the first and fourteenth iteration. Squares in the upper and the lower triangle indicate V_{upij} and V_{downij} respectively. The linear dimension of each rectangle is proportional to the value of ΔU_{upij}, ΔU_{downij}, U_{upij}, U_{downij}, V_{upij}, and V_{downij}. Black and white rectangles indicate positive and negative values respectively. Fig. 4-6a shows the intermediate state of 44 neurons after the first iteration where 7 edges are embedded but some edges intersect each other. Fig. 4-6b shows the final state of 44 neurons after the fourteenth iteration where 20 edges are embedded and no edges intersect each other. Our simulator discovered that the new maximal planar subgraph contains 20 edges instead of 19 edges, which contradicts to the result of Jayakumar.

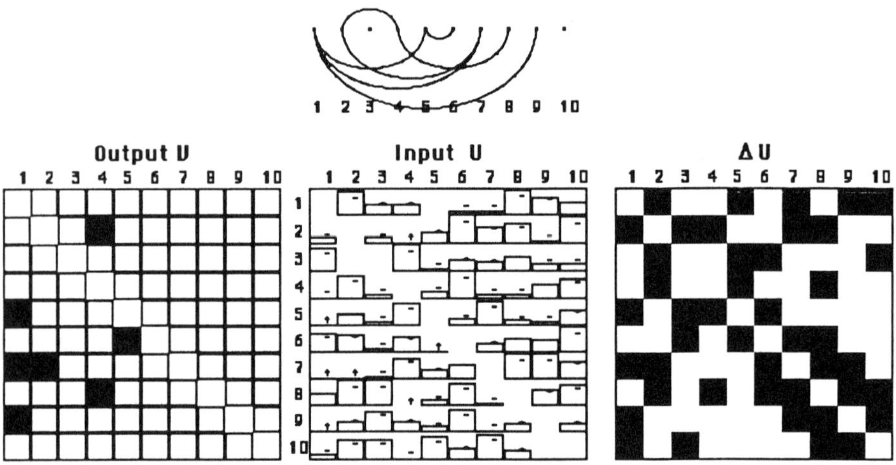

Fig. 4-6 (a) The convergence of the graph planarization neural network to a solution

Fig. 4-7 shows the simulation result where several sets of the coefficients were experimented. It indicates that either 20 edges or 19 edges out of 22 edges can be consistently embedded in a single plane.

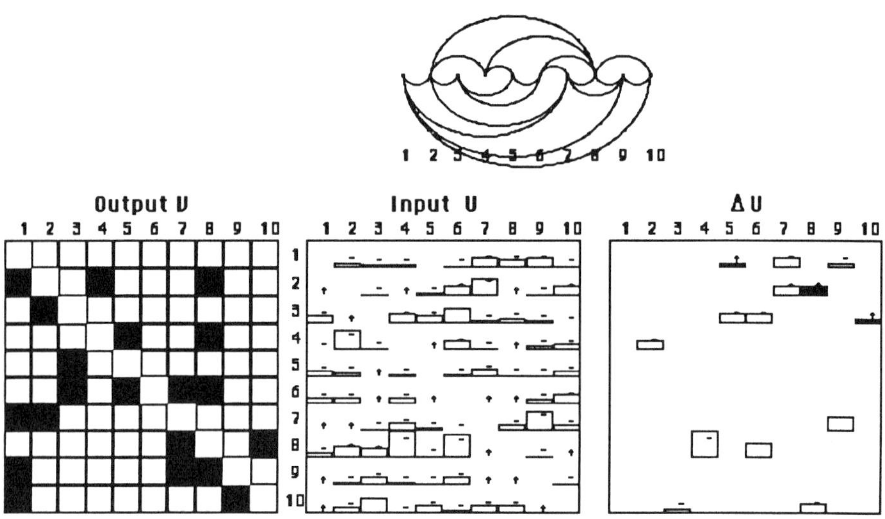

Fig. 4-6 (b) The convergence of the graph planarization neural network to a solution

Coefficients		The number of iteration steps	The number of embedded edges
A	B		
2.5	1	6	20
2	1	9	20
1.5	1	7	20
1.25	1	13	20
1	1	9	19
1	1.5	9	19
1	2	9	19
1	2.5	9	19

Fig. 4-7 The simulation result on the relationship between the coefficients, the number of iteration steps, and the number of embedded edges

Hill-Climbing Terms

We have further experimented the middle size graph problem with 48 vertices and 105 edges as shown in Fig. 4-8a. Although the graph in Fig. 4-8a looks a sophisticated graph, the graph is given by the US-continental map where it contains 48 (states) vertices and 105 edges and which is known to be planar.

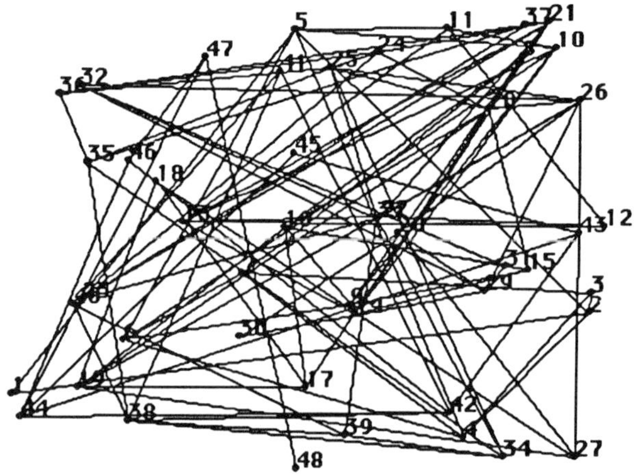

Fig. 4-8 (a) Graph with 48 vertices and 105 edges

The graph with 48 vertices and 105 edges in the single row representation is shown in Fig. 4-8b. With the same set of the coefficients and the unit time, the state of the system converged to the near-global minimum within 20 iteration steps. The near global solutions contain about 100 edges.

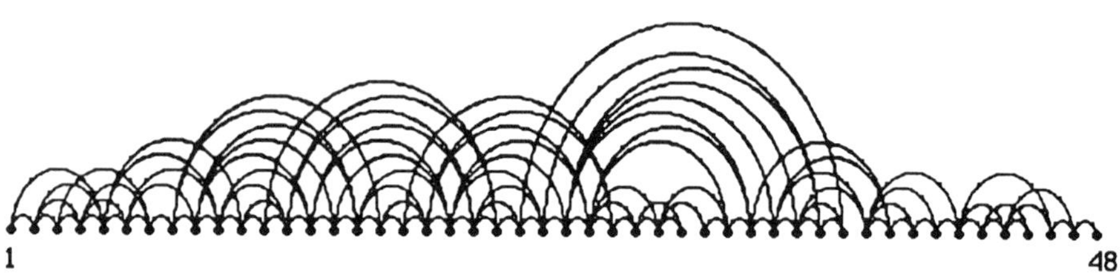

Fig. 4-8 (b) Graph in the single-row representation with 48 vertices and 105 edges

In order to improve the solution quality, the following hill-climbing term was introduced: $+Ch(Vup_{ij}+Vdown_{ij})$ where $h(x)=1$ if $x=0$, 0 otherwise. The new term was added to Eq. (4-1) and Eq. (4-2). Only if the state of two neurons is ($Vup_{ij}=0$&$Vdown_{ij}=0$), this term will perform the excitatory force else it will take no action.

When A=B=1, C=5, Dt=10^{-6}, and the initial values of $Uup_{ij}(t)$ and $Udown_{ij}(t)$ were randomized in a range of 0 to -1/2000, the state of the system converged to the global minimum in the 206th iteration. The solution is depicted in Fig. 4-8c. Several experiments using different sets of the coefficients showed the consistent improvement in the solutions. We also changed the numbering of state regions, our system produced the planar subgraph with more than 100 edges in 20 iteration steps.

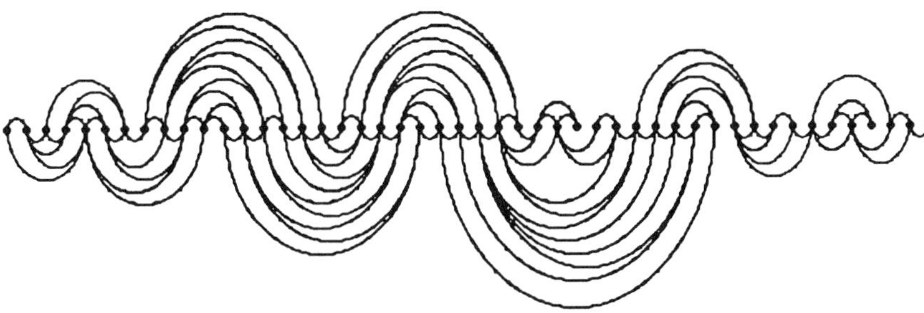

Fig. 4-8 (c) Solution of the graph in Fig. 4-8b

Based on the proposed parallel planarization algorithm we have developed the simulator. The developed simulator is currently running on a Macintosh and on an HP Apollo 3500 computer. Within O(1) time the algorithm not only generates a near-optimum planar subgraph from the nonplanar or planar graph but also embeds the subgraph on a plane. The algorithm can be implemented by an N x N 2-dimensional neural network array where N is the number of vertices in a given graph. Among N^2 neurons only 2M neurons are used to obtain the solution where M is the number of edges in the graph. Remember that the number of required neurons for M-edge graph planarization is actually less than 2M where it is actually determined by the number of intersected edges. As far as we have observed the behavior of the simulator the state of the system always converges to the good solution within 20 or 30 iteration steps.

4.3 REFERENCES

Garey M. R., and Johnson D. S., (1979), Computers and Intractability: A Guide to the theory of NP-completeness, (W. H. Freeman and Company).

Jayakumar R., Thulasiraman K., and Swamy M. N. S., (March 1989), $O(n^2)$ algorithms for graph planarization, IEEE Trans. on Computer Aided Design, 8, 3.

Tarjan R. E., (1971), An efficient planarity algorithm, STAN-CS-244-71, Computer Science Dept., Stanford University.

Takefuji Y., and Lee K. C., (1989), A near-optimum parallel planarization algorithm, Science, 245, 1221-1223.

4.4 EXERCISES

1. How many neurons are needed to solve the graph in Fig. 4-4.
2. Build a simulator to solve the graph in Fig. 4-4 and observe the behavior of your system by changing the coefficients and the unit time. Find the global optimum solution by exhaustive search and compare it with the result of your simulator.
3. Add the hill-climbing term in your simulator to observe the effect.
4. Generate several different size graphs and observe the number of iteration steps to obtain the local minimum solution using your simulator.
5. Discuss the termination condition of your system or the condition of the equilibrium state. How often does your system show the oscillatory behavior?
6. Survey algorithms for graph planarization and graph embedding. Compare the solution quality and the computation time of the neural network system with the best known algorithm.
7. The solution quality is influenced by vertex-ordering. Observe the effect of vertex-ordering by assigning the different numbers to the vertices in Fig. 4-4. Develop the algorithm for vertex-ordering.
8. Build a simulator to solve the graph in Fig. 4-4 where all of computations are based on integers and no floating number is used.

Chapter 5

CHANNEL ROUTING PROBLEMS

A parallel algorithm for channel routing problems is introduced in this Chapter. The channel routing problem is very important in automatic layout design of VLSI circuits and printed circuit boards. The problem is to route the given interconnections between two rows of terminals on a multi-layer channel where the channel area must be minimized. Although several algorithms have been reported for two-layer problems, two-layer-and-over-the-cell problems, and three-layer problems, the current advancement of the VLSI technology allows us to use four layers composed of two metal layers and two polysilicon layers for routing in a chip where no parallel algorithms have been proposed. The goal of the proposed parallel algorithm is to find a near-optimum routing solution for the given interconnections in a short time. The algorithm is used for the four-layer channel routing problem where it requires $n \times m \times 2$ processing elements (neurons) for the n-net-m-track problem. The algorithm has three advantages over the conventional algorithms; 1) it can be easily modified for accommodating more than four-layer channel routing problems, 2) it runs not only on a sequential machine but also on a parallel machine with maximally $n \times m \times 2$ processors, and 3) the program size is very small. The algorithm is tested by solving seven benchmark problems where the algorithm finds routing solutions in nearly constant time with $n \times m \times 2$ processors. This Chapter is based on a paper published in IEEE Trans. on Computer-Aided Design (Funabiki and Takefuji 1991).

5.1 INTRODUCTION

The channel routing problem in a multi-layer channel is very important in automatic layout design of VLSI circuits and printed circuit boards. A channel consists of two parallel, horizontal rows of points which are called terminals. The terminals are placed at regular intervals and identify the columns of the channel. A net consists of a set of terminals that must be interconnected through some routing paths. Some nets may have a connection point at one or both ends (top and/or bottom) of the channel. The channel routing problem is not only to route the given nets or interconnections between the terminals on the multi-layer channel, but also to minimize the channel area.

The sequential algorithms for the two-layer channel routing problems have been extensively studied by many researchers (Hashimoto A. and Stevens J. 1971) (Kernighan B. W., Schweikert D. G., and Persky G. 1973) (Deutsch D. N. 1976) (Sahni S. and Bhatt A. 1980) (LaPaugh A. S. 1980) (Dolev D. et al. 1981) (Yoshimura T., and Kuh E. S. 1982) (Rivest R. L. and Fiduccia C. M. 1982) (Leong H. W. and Liu C. L. 1983) (Burstein M. and Pelavin R. 1983a) (Burstein M. and Pelavin R. 1983b) (Szymanski T. G. 1985) (Reed J., Sangiovanni-Vincentelli A., and Santomauro M. 1985) (Leong H. W., Wong D. F., and Liu C. L. 1985) (Joobbani R. and Siewiorek D. P. 1985). The existing algorithms have the following common features. The routing paths consist of the horizontal segments which are parallel to the terminals of the channel and the vertical segments. All the horizontal segments of the routing paths are assigned on one layer and all the vertical segments of them are assigned on another layer. The connections between the horizontal segments and the vertical segments are made through the contact windows which are called via holes. For the integrated circuits, typically the horizontal segments are embedded on a metal layer while the vertical segments are embedded on a polysilicon and/or diffusion layer. Any two routing paths on the same layer cannot be placed within some distance of each other which is called separation condition. For convenience, a unit grid is superimposed on the channel where the size of one unit satisfies the separation condition and all the terminals are located at the grid points. All the routing paths on the channel must follow the grid lines. The horizontal segments are called tracks and the vertical segments are called columns. In this model, the separation condition is that any two nets must be embedded neither on the same track nor on the same column if they overlap there, which is called overlapping condition. When the width of the channel is fixed, minimizing the channel area is equivalent to minimizing the number of required tracks where all the given nets must be embedded.

In some sequential algorithms, doglegging is introduced where a routing path of a net is split into more than two horizontal segments on different tracks (Deutsch D. N. 1976) (Yoshimura T. and Kuh E. S. 1982) (Rivest R. L. and Fiduccia C. M. 1982) (Leong H. W. and Liu C. L. 1983) (Burstein M. and Pelavin R. 1983a) (Burstein M. and Pelavin R. 1983b) (Szymanski T. G. 1985) (Reed J., Sangiovanni-Vincentelli A., and Santomauro M. 1985) (Leong H. W., Wong D. F., and Liu C. L. 1985) (Joobbani R. and Siewiorek D. P. 1985). Doglegging is sometimes effective to reduce the number of tracks of the channel and to solve the cyclic conflict. The cyclic conflict occurs when one net interconnects a top terminal on the i-th column with a bottom terminal on the j-th column while another net interconnects a bottom terminal on the i-th column with a top terminal on the j-th column. Note that a top terminal means a terminal at the top end of a channel and a bottom terminal means a terminal at the bottom end of a channel. However, this cyclic conflict occurs infrequently and it can often be avoided by rearranging the terminal placement. Doglegging requires additional via holes which reduce the reliability of the VLSI system and increase the manufacturing cost. It is desirable either to reduce the number of dogleggs or to eliminate dogleggs completely for the practical use.

To further reduce the channel area, several sequential algorithms for two-layer-and-over-the-cell channel routing problems (Deutsch D. N. and P. Glick 1980) (Krohn H. E. 1983) (Shiraishi Y. and J. Sakemi J. 1987) (Gudmundsson G. and Ntafos S. 1987) (Cong J. and Liu C. L. 1988) (Cong J. and Liu C. L. 1990) and for three-layer channel routing problems have been also proposed (Chen Y. K. and M. L.

Liu 1984) (Cong J., Wong D. F., and Liu C. L. 1987). The over-the-cell channel routing algorithms use not only the conventional two layers of channel but also the area over the cells for interconnections. The three-layer channel routing algorithms use two layers for the horizontal segments and one layer for the vertical segments. Although these algorithms find better solutions than the algorithms for two-layer problems, they are still based on two-layer routing technique so that it is not easy to apply them for more than three layers problems. It is not guaranteed to find the optimum solution with their algorithms.

The channel routing problems are proved to be NP-complete. LaPaugh shows the NP-completeness of the channel routing problem without doglegging (LaPaugh 1980). Szymanski shows the NP-completeness of the channel routing problem with doglegging (Szymanski 1985). Gudmundsson et al. shows the NP-completeness of the over-the-cell channel routing problem (Gudmumdsson and Ntafos 1987).

Several parallel routing algorithms have been proposed for the knock-knee model (Chang and JaJa 1988a) (Krishnamurthy and JaJa 1988), for the river routing (Chang and JaJa 1988b)(Chang and JaJa 1988c), and for the global routing (Iosupovici 1986)(Rose 1990). A parallel channel routing algorithm based on simulated annealing was also proposed in (Brouwer and Banerjee 1988), but the algorithm does not guarantee to find optimum solutions. Although parallel channel routing algorithms which flexibly accommodate the future advancement of the VLSI technology have been highly demanded, a few parallel algorithms have been reported.

5.2 GRAPH PLANARIZATION AND CHANNEL ROUTING

The current advancement of the VLSI technology allows us to use four layers composed of two metal layers and two polysilicon layers for routing in a chip (Mosis 1990). The algorithm embeds the given nets on one of the two independent two-layer channels which have the same structure as the conventional two-layer channels. In other words, the four-layer channel has two layers for the horizontal segments of the nets and two layers for the vertical segments of them. The proposed algorithm with slight modification is able to solve two-layer problems, two-layer-and-over-the-cell problems, and three-layer problems.

The proposed parallel algorithm is based on the three dimensional artificial neural network model which is composed of a large number of massively interconnected simple processing elements (neurons).

The output V_{ijk} of the ijk-th processing element based on the modified McCulloch-Pitts neuron model follows:

$V_{ijk}=1$ if $U_{ijk}>0$ and $U_{ijk}=\max\{ U_{iqr} \}$ for q=1,...,m and r=1,2.

$=0$ otherwise. (5.1)

where U_{ijk} is the input of the ijk-th processing element and m is the number of tracks of the channel.

Fig. 5-1a shows a channel routing problem in (Yoshimura and Kuh 1982) where ten nets are given to be routed in a four-layer channel which has three tracks. Ten nets are (T2,T5), (B1,B6), (B2,B4), (T3,T9), (B3,T4,B5), (T6,B7), (T7,B11), (B8,B10), (B9,T10,B12), and (T11,T12) where Ti indicates the top terminal at the i-th column and Bj indicates the bottom terminal at the j-th column.

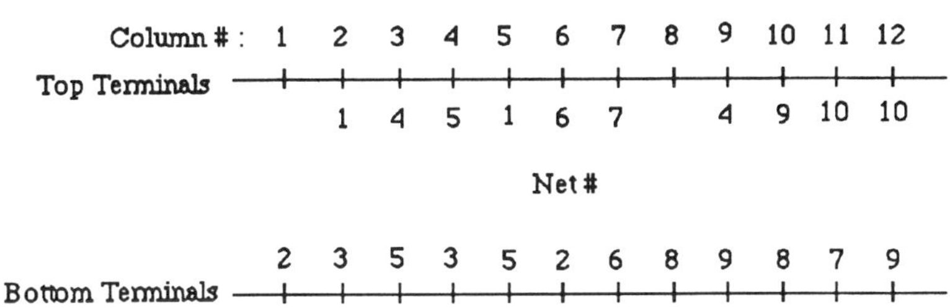

Fig. 5-1 (a) 10-net-3-track problem

For example, #1 net (T2,T5) has two terminals to be interconnected; one is the top terminal at the second column and another is the top terminal at the fifth column.

Fig. 5-1b shows the system representation for this problem. Only one horizontal segment is used for one net, and the vertical segments of the net are automatically assigned on the layer corresponding to the layer on which the horizontal segment of the net is assigned. The channel routing problem can be simplified into the layer-track problem. It is to find the layer number and the track number for embedding the horizontal segment of the given net without violating the overlapping conditions.

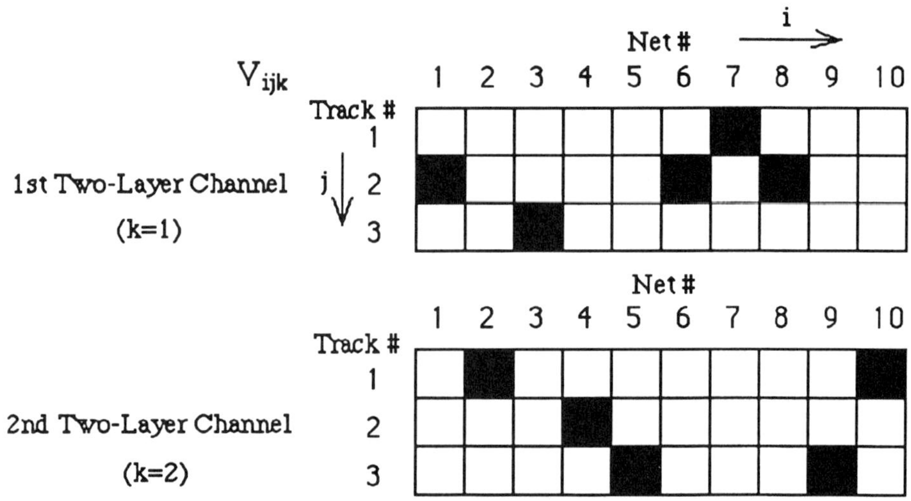

Fig. 5-1 (b) The output state of processing elements

Six(=3×2) processing elements (neurons) are used to indicate on which layer and track one given net should be embedded. Because three tracks of two layers for the horizontal segments are available for the ten-net and three-track problem, the total of 60(=10×3×2) processing elements are used in this problem. V_{ijk} represents the output of the ijk-th processing element which is corresponding to the i-th net and the j-th track on the k-th layer for i=1,...,10, j=1,2,3, and k=1,2. Generally n×m×2 processing elements are used to represent n nets, m tracks, and two horizontal segment layers for the four-layer channel routing problems where m×2 processing elements describe the track number and the layer number for a single net.

The output of one and only one processing element among the m×2 processing elements should be nonzero to locate the net on one of the m tracks of two layers. The nonzero output means that the net should be embedded on the corresponding track and layer. Fig. 5-1b shows one of the solutions for this problem where the black squares indicate the nonzero output of the processing elements and the white squares indicate the zero output of the processing elements. Fig. 5-1b shows that the

#1 net (T2,T5) is assigned on the second track of the first two-layer channel and the #2 net (B1,B6) is assigned on the first track of the second two-layer channel, and so on. Note that the two-layer channel means a pair of two layers for the horizontal segments and vertical segments. Fig. 5-1c shows the routing solution corresponding to Fig. 5-1b.

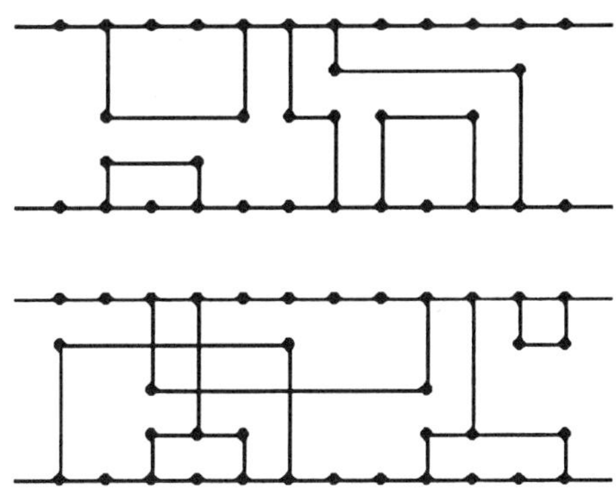

Fig. 5-1 (c) The routing solution corresponding to Fig. 5-1b

Each net must satisfy the separation conditions, in other words, any two different nets must not violate the overlapping conditions. Fig. 5-2 shows the overlapping conditions for the horizontal segments of the nets where $head_i$ indicates the left-most column number of the i-th net and $tail_i$ indicates the right-most column number of the i-th net. The horizontal overlapping conditions for the i-th-net-j-th-track-k-th-layer processing element are given by:

$$\sum_{\substack{p=1 \\ p\neq i \\ head_i \leq head_p \leq tail_i}}^{n} V_{pjk} + \sum_{\substack{p=1 \\ p\neq i \\ head_p \leq head_i \leq tail_p}}^{n} V_{pjk} \qquad (5.2)$$

This horizontal condition is nonzero if the horizontal segments of the other nets overlap the horizontal segment of the i-th net on the j-th track of the k-th layer.

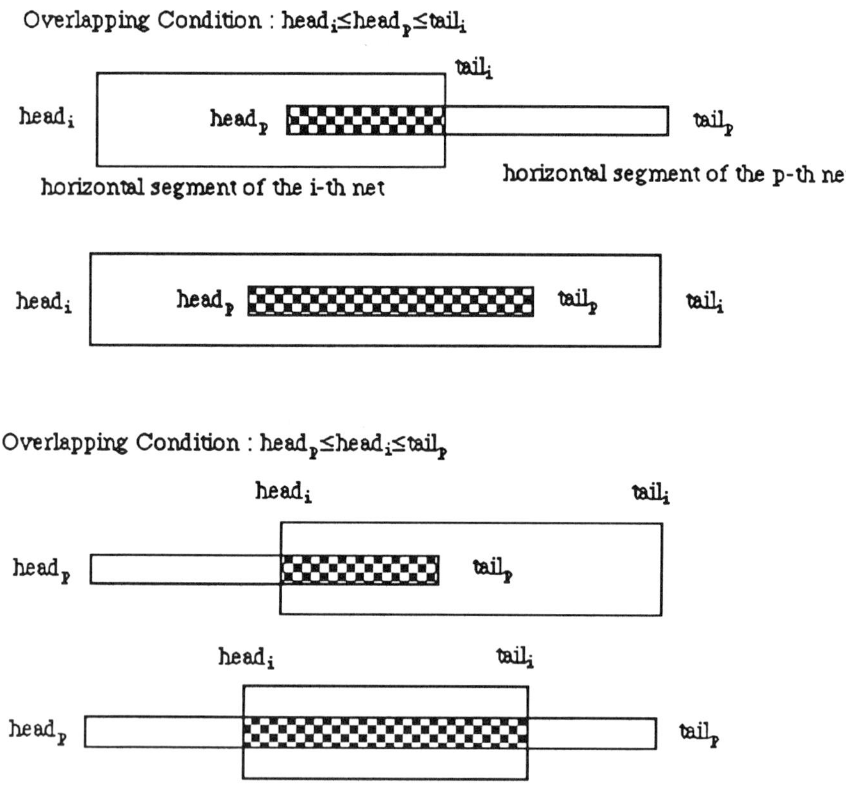

Fig. 5-2 Overlapping conditions for horizontal segments

Fig. 5-3 shows the overlapping conditions for the vertical segments of the nets where the i-th net and the p-th net have the terminals on the opposite sides of the same column. The vertical overlapping conditions for the i-th-net-j-th-track-k-th-layer processing element are given by:

$$\sum_{\substack{p=1 \\ p \neq i}}^{n} T_{ip} \sum_{q=1}^{j} V_{pqk} + \sum_{\substack{p=1 \\ p \neq i}}^{n} B_{ip} \sum_{q=j}^{m} V_{pqk} \tag{5.3}$$

where T_{ip} is 1 if the p-th net has a bottom terminal at the column at which the i-th net has a top terminal, 0 otherwise, and B_{ip} is 1 if the p-th net has a top terminal at the column at which the i-th net has a bottom terminal, 0 otherwise. The vertical condition is nonzero if the vertical segments of the other nets overlap the vertical segments of the i-th net.

The motion equation of the i-th-net-j-th-track-k-th-layer processing element for the n-net-m-track problem is given by:

$$\frac{dU_{ijk}}{dt} = -A\left(\sum_{q=1}^{m}\sum_{r=1}^{2} V_{iqr} - 1\right) - B\left(\sum_{\substack{p=1\\p\neq i\\head_i \leq head_p \leq tail_i}}^{n} V_{pjk} + \sum_{\substack{p=1\\p\neq i\\head_p \leq head_i \leq tail_p}}^{n} V_{pjk}\right)$$

$$-B\left(\sum_{\substack{p=1\\p\neq i}}^{n} T_{ip}\sum_{q=1}^{j} V_{pqk} + \sum_{\substack{p=1\\p\neq i}}^{n} B_{ip}\sum_{q=j}^{m} V_{pqk}\right) + Ch\left(\sum_{q=1}^{m}\sum_{r=1}^{2} V_{iqr}\right) \qquad (5.4)$$

The first term (A-term) in Eq. (5.4) forces one and only one output among the 2m processing elements to be nonzero where the i-th net is assigned. The second and third terms (B-terms) perform the inhibitory forces. The B-terms discourage the output of the ijk-th processing element to be nonzero if the other nets overlap with the i-th net. The last term (C-term) provides the hill-climbing which allows the state of the system to escape from the local minimum and to converge to the global minimum. The C-term encourages the output of the ijk-th processing element to be nonzero if the output of all the processing elements for the i-th net is zero. The function h(x) is 1 if x=0, 0 otherwise. A, B, and C are constant coefficients.

The simulator was developed based on Program2 in Chapter 1 as shown in Fig. 1-8. The following parameters are used: t=0, A=B=1,C=10, U_max=20, U_min=-20. B-terms in Eq. (5.4) are alternatively used: if (t mod 10) < 5 then the B-terms in Eq. (5.4) are given by:

$$-B\Big[\sum_{\substack{p=1\\p\neq i\\head_i\leq head_p\leq tail_i}}^{n} V_{pjk}(t)+\sum_{\substack{p=1\\p\neq i\\head_p\leq head_i\leq tail_p}}^{n} V_{pjk}(t)\Big]V_{ijk}(t) \text{ and}$$

$$-B\Big[\sum_{\substack{p=1\\p\neq i}}^{n} T_{ip}\sum_{q=1}^{j} V_{pqk}(t)+\sum_{\substack{p=1\\p\neq i}}^{n} B_{ip}\sum_{q=j}^{m} V_{pqk}(t)\Big]V_{ijk}(t)$$

else they are given by:

$$-B\Big[\sum_{\substack{p=1\\p\neq i\\head_i\leq head_p\leq tail_i}}^{n} V_{pjk}(t)+\sum_{\substack{p=1\\p\neq i\\head_p\leq head_i\leq tail_p}}^{n} V_{pjk}(t)\Big] \text{ and}$$

$$-B\Big[\sum_{\substack{p=1\\p\neq i}}^{n} T_{ip}\sum_{q=1}^{j} V_{pqk}(t)+\sum_{\substack{p=1\\p\neq i}}^{n} B_{ip}\sum_{q=j}^{m} V_{pqk}(t)\Big].$$

It is also important that the values of the inputs are always kept in the certain range where they are computed by: If $U_{ijk}(t+1)$>U_max then $U_{ijk}(t+1)$=U_max and If $U_{ijk}(t+1)$<U_min then $U_{ijk}(t+1)$=U_min for i=1,...,n, j=1,...,m, and k=1,2 where U_max and U_min are the upper and lower limits of $U_{ijk}(t+1)$ respectively.

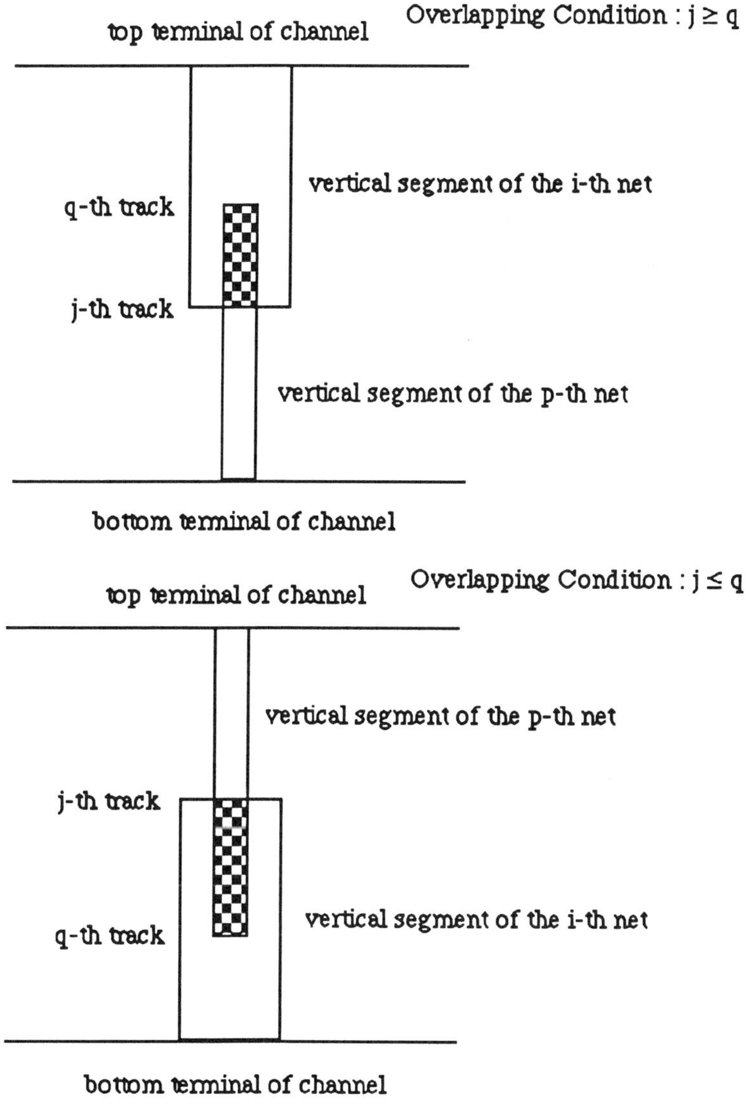

Fig 5-3 Overlapping conditions for vertical segments

The termination condition is given by: If $V_{ijk}(t)=1$ and $\Delta U_{ijk}(t)=0$ for $i=1,...,n$,

$\exists j \in \{1,...,m\}$, and $\exists k \in \{1,2\}$ (all nets are embedded without conflicts) or t=T_max then terminate Program2. T_max is the maximum number of iteration steps.

We divide the given nets into two groups to perform this algorithm in two phases. The first group consists of the longer nets where the net width between the left-most column and the right-most column is more than 30 % of the total channel width, and the other group consists of the remaining shorter nets. It is necessary to fix the locations of the first group nets in the beginning. Because it is difficult or impossible to find the locations for the longer nets in the channel if many nets are embedded. Our strategy is that in the first phase we assign the first group nets by applying our algorithm only for them and in the second phase we assign the second group nets by applying the algorithm for all the nets where the output of the processing elements corresponding to the first group nets are fixed.

The state of $n \times m \times 2$ processing elements for the n-net-m-track problem can be updated synchronously or asynchronously. The developed simulator solved seven benchmark problems in (Yoshimura and Kuh 1982) where it took less than 10 minutes for one convergence on Macintosh IIfx. The benchmark problems have no cyclic conflict. Fig. 5-4 and 5-5 show solutions of our algorithm for two of seven problems respectively. Our algorithm found several other solutions in the same problems from different initial values of $U_{ijk}(t)$. Table 5-1 shows the numbers of nets in seven problems and the numbers of tracks which our algorithm needs to route all nets in the four-layer channel. They are compared with the two-layer routing solutions in (Yoshimura and Kuh 1982) and the three-layer routing solutions in (Chen and Liu 1984) where no doglegging is allowed. Our algorithm found the theoretically optimum solutions for the benchmark problems except the Deutsch's example where it found the near-optimum solutions.

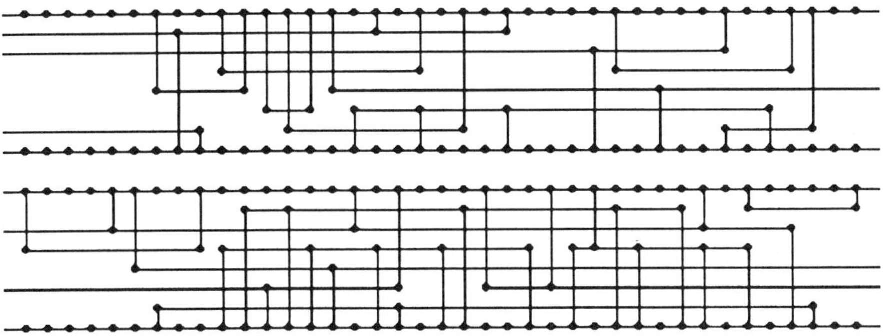

Fig. 5-4 One of solutions for the example 1

Table 5-2 shows the frequency of the convergence to the optimum solutions and the average numbers of iteration steps where 100 simulation runs were performed for each problem. For each simulation run, different initial values of $U_{ijk}(t)$ are randomly generated. Fig. 5-6 shows the relationship between the frequency and the number of iteration steps to converge to the optimum solutions in two problems. The simulation results empirically show that our algorithm solves the channel routing problems in a nearly constant time with $n \times m \times 2$ processors. It is observed that the O(nm) time sequential algorithm is achieved by the modified proposed algorithm where n is the number of nets on an m-track-four-layer channel.

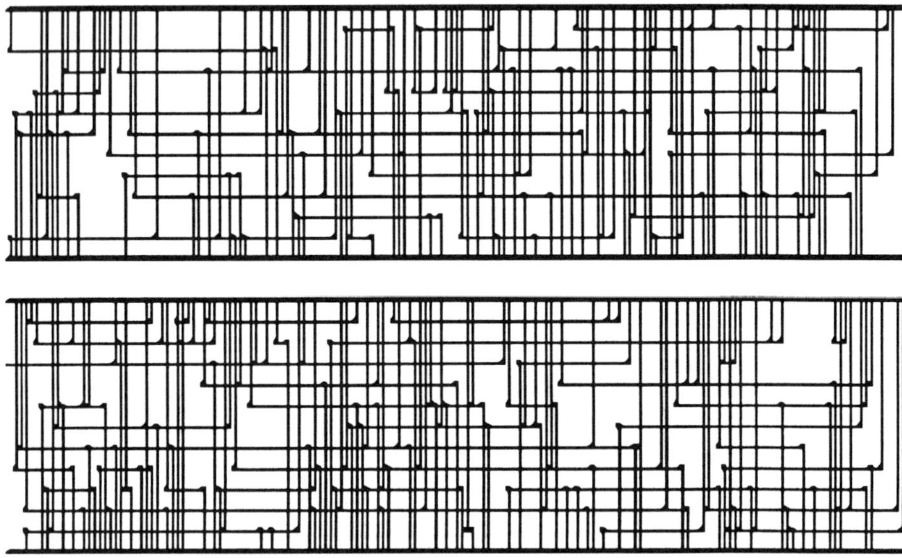

Fig. 5-5 One of solutions for Deutsch's difficult example

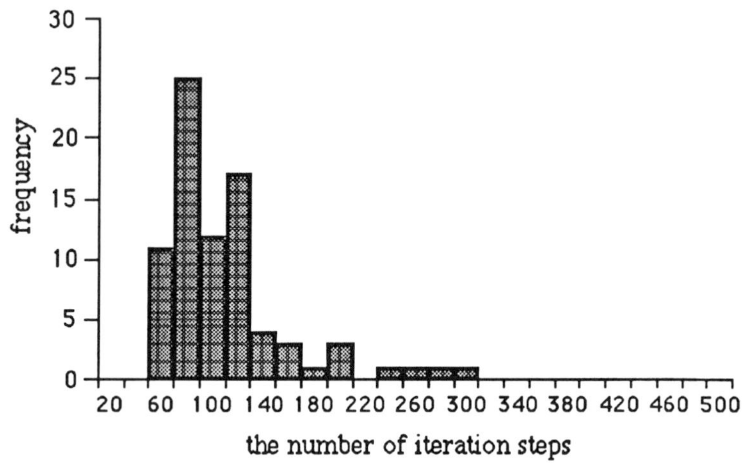

Fig. 5-6 Relationship between the number of iteration steps and the frequency to converge to optimum solutions in the example 5

Table 5-1 Comparisons of track numbers with other no-doglegging routers

problem #	number of nets	our six-layer solutions	two-layer solutions by Yoshimura & Kuh	three-layer solutions by Chen & Liu
the example 1	21	3	12	7
the example 3a	45	4	15	8
the example 3b	47	5	17	10
the example 3c	54	5	18	9
the example 4b	55	5	17	13
the example 5	61	5	20	10
Deutsch's difficult example	72	7	28	23

Table 2 Summary of simulation results

Problem No.	Average Iteration Steps to Optimum Solutions	Convergence Frequency to Optimum Solutions
Example 1	84.4	44%
Example 3a	147.6	29%
Example 3b	89.9	53%
Example 3c	280.4	7%
Example 4b	150.3	40%
Example 5	101.5	80%
Difficult Example	133.1	17%

Modifications for More Than Four-Layer Problems

In order to show the flexibility of the proposed parallel algorithm, we modified and applied the algorithm for 2s-layer channel routing problems where a channel has s layers for the horizontal segments of the nets and s layers for the vertical segments of them. The motion equation in Eq. (5.4) is slightly modified as follows:

$$\frac{dU_{ijk}}{dt}=-A\left(\sum_{q=1}^{m}\sum_{r=1}^{s}V_{iqr}-1\right)-B\left(\sum_{\substack{p=1\\p\neq i\\head_i\leq head_p\leq tail_i}}^{n}V_{pjk}+\sum_{\substack{p=1\\p\neq i\\head_p\leq head_i\leq tail_p}}^{n}V_{pjk}\right)$$

$$-B\left(\sum_{\substack{p=1\\p\neq i}}^{n}T_{ip}\sum_{q=1}^{j}V_{pqk}+\sum_{\substack{p=1\\p\neq i}}^{n}B_{ip}\sum_{q=j}^{m}V_{pqk}\right)+Ch\left(\sum_{q=1}^{m}\sum_{r=1}^{s}V_{iqr}\right) \qquad (5.5)$$

As the example, we examined Deutsch's difficult example in six-layer and eight-layer problems. Fig. 5-7 and 5-8 show solutions of six-layer and eight-layer problems respectively. Table 5-3 shows the numbers of tracks which the algorithm needs to route all nets in the 4-layer through 10-layer channel for Deutsch's difficult example. Our algorithm found the optimum solutions for each problem in this sHsV (s-horizontal-layer-s-vertical-layer) model. The algorithm can be also easily modified and extended for solving the 3s-layer channel routing problems in the 2sHsV model. The 2sHsV model has the possibility to give less track solutions in the same problem than the sHsV model. We conclude that our parallel algorithm is so flexible that it can easily accommodate the future advancement of the silicon technology.

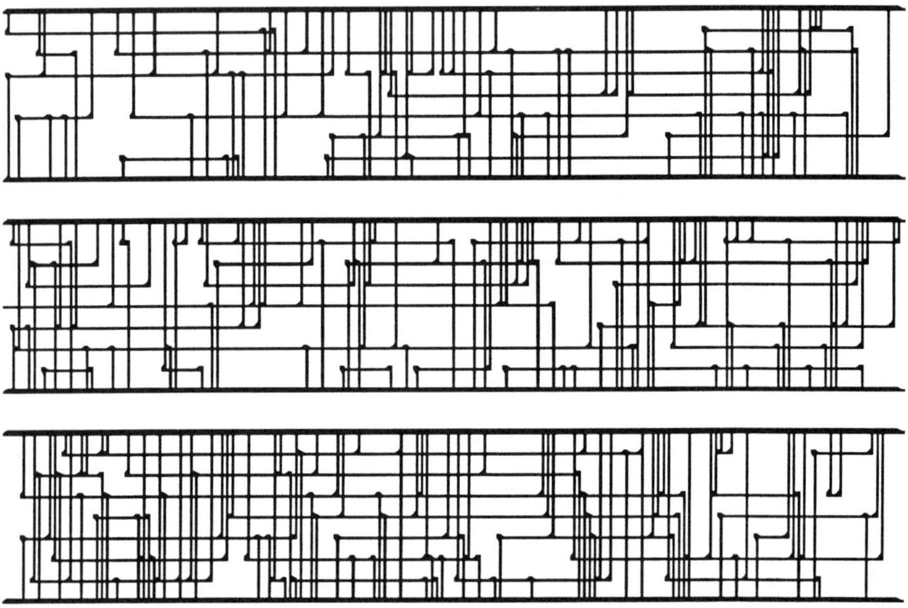

Fig. 5-7 Six-layer solution of Deutsch's difficult example

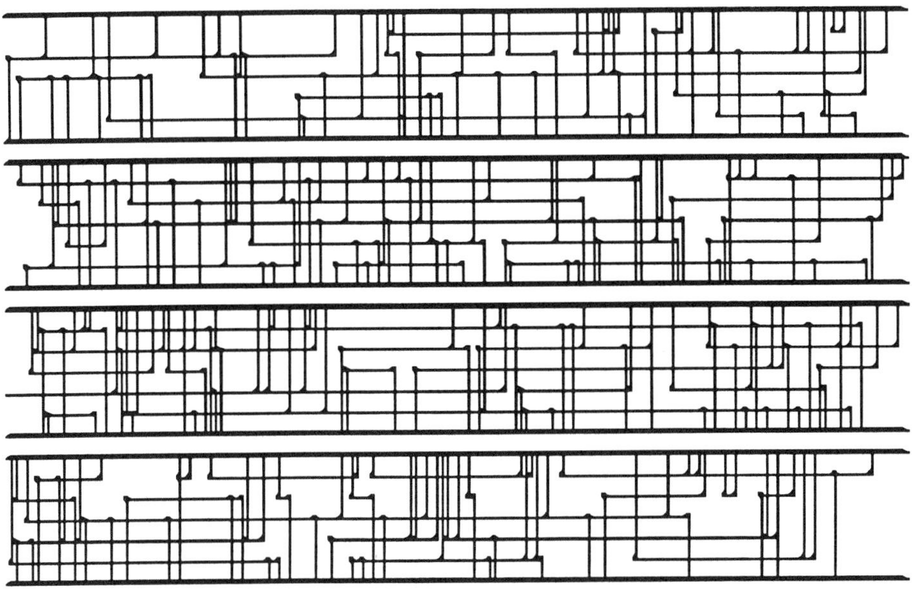

Fig. 5-8 Eight-layer solution of Deutsch's difficult example

Table 5-3 Track numbers in 4-10 layer solutions of Deutsch's difficult example

Number of Layers	Number of Tracks
4	11
6	7
8	5
10	4

This Chapter proposes the parallel algorithm for the four-layer channel routing problems and the 2s-layer channel routing problems in the sHsV model. The algorithm requires $n \times m \times 2$ processing elements for the n-net-m-track routing problem on the four-layer channel. The algorithm runs not only on a sequential machine but also on a parallel machine with maximally $n \times m \times 2$ processors. In seven benchmark problems, the algorithm finds routing solutions in nearly constant time with $n \times m \times 2$ processors. The simulation results supports the consistency of the algorithm. They also show that the primary goal of finding the near-optimum solutions in parallel processing is successfully achieved in terms of the computation time and the solution quality. The algorithm can be easily modified and extended for solving the 3s-layer channel routing problems in the 2sHsV model. It might be interesting to investigate a parallel algorithm using doglegging.

5.3 REFERENCES

Brouwer R. J. and Banerjee P., (1988), A parallel simulated annealing algorithm for channel routing on a hypercube multiprocessor, IEEE Int. Conf. Comput. Design, pp. 4-7.

Burstein M., and Pelavin R., (1983a), Hierarchical channel router, Proc. 20th Design Automation Conf., pp. 591-597.

Burstein M., and Pelavin R., (1983b), Hierarchical wire routing, IEEE Trans. CAD, Vol. CAD-2, 4, pp. 223-234.

Chang S. C. and JàJà J., (1988a), Parallel algorithms for channel routing in the knock-knee model, Proc. Int. Conf. Parallel Processing, pp. 18-25.

Chang S. C. and JàJà J., (1988b), Parallel algorithms for river routing, Proc. Int. Conf. Parallel Processing, pp. 9-13.

Chang S. C. and JàJà J., (1988c), Optimal mesh algorithms for VLSI routing, Proc. 2nd Symp. Frontiers Massively Parallel Comput., pp. 125-128.

Chen Y. K. and M. L. Liu, (1984), Three-layer channel routing," IEEE Trans. CAD, Vol. CAD-3, 2, pp. 156-163.

Cong J. and Liu C. L., (1988), Over-the-cell channel routing, Proc. ICCAD-88, pp. 80-83.

Cong J. and Liu C. L., (1990), Over-the-cell channel routing, IEEE Trans. CAD, Vol. 9, 4, pp. 408-418.

Cong J., Wong D. F., and Liu C. L., (1987), A new approach to the three layer channel routing problem," Proc. ICCAD-87, pp. 378-381.

Deutsch D. N., (1976), A dogleg channel router, Proc. 13th Design Automation Conf., 425-433.

Deutsch D. N. and P. Glick, (1980), An over-the-cell router, Proc. 17th IEEE/ACM Design Automation Conf., pp. 32-39.

Dolev D. et al., (1981), Optimal wiring between rectangles, Proc. 13th Annual ACM Symp. Theory of Computing, 312-317.

Funabiki N. and Takefuji Y., (1991) A parallel algorithm for channel routing problems, IEEE Trans. on CAD, 10, 12.

Gudmundsson G. and Ntafos S., (1987), Channel routing with superterminals, Proc. 25th Allerton Conf. on Computing, Control and Communication, pp. 375-376.

Hashimoto A., and Stevens J., (1971), Wire routing by optimizing channel assignment with large apertures, Proc. 8th Design Automation Workshop, 155-169.

Iosupovici A., (April 1986), A class of array architectures for hardware grid routers, IEEE Trans. CAD, Vol. CAD-5, pp. 245-255.

Joobbani R., and Siewiorek D. P., (1985), Weaver: A knowledge-based routing expert, Proc. 22nd Design Automation Conf., pp. 266-272.

Kernighan B. W., Schweikert D. G., and Persky G., (1973), An optimum channel-routing algorithm for polycell layouts of integrated circuits, Proc. 10th Design Automation Workshop, 50-59.

Krishnamurthy S. and J. JàJà, (1988), Provably good parallel algorithms for channel routing of multi-terminal nets," Proc. 2nd Symp. Frontiers Massively Parallel Comput., pp. 177-180.

Krohn H. E., (1983), An over-cell gate array channel router, Proc. 20th Design Automation Conf., pp. 665-670.

LaPaugh A. S., (1980), Algorithms for integrated circuits layout: An analytic approach, Ph.D. dissertation, M.I.T. Lab. Computer Science.

Leong H. W., and Liu C. L., (1983), A new channel routing problem, Proc. 20th Design Automation Conf., pp. 584-590.

Leong H. W., Wong D. F., and Liu C. L., (1985), A simulated-annealing channel router, Proc. ICCAD-85, pp. 226-228.

Mosis user manual, supplements, (1990).

Reed J., Sangiovanni-Vincentelli A., and Santomauro M., (July 1985), A new symbolic channel router: YACR2, IEEE Trans. CAD, Vol. CAD-4, 3, pp. 208-219.

Rivest R. L., and Fiduccia C. M., (1982), A greedy channel router, Proc. 19th Design Automation Conf., pp. 418-424.

Rose J., (Oct. 1990), Parallel global routing for standard cells, IEEE Trans. CAD, Vol. 9, 10, pp. 1085-1095.

Sahni S., and Bhatt A., (1980), The complexity of design automation problems, Proc. 17th Design Automation Conf., 402-411.

Shiraishi Y. and J. Sakemi J., (1987), A permeation router, IEEE Trans. CAD, Vol. CAD-6, 3, pp. 462-471.

Szymanski T. G., (Jan. 1985), Dogleg channel routing is NP-complete, IEEE Trans. CAD, Vol. CAD-4, 1, pp. 31-41.

Yoshimura T., and Kuh E. S., (Jan. 1982), Efficient algorithms for channel routing, IEEE Trans. CAD, Vol. CAD-1, 1, pp. 25-35.

5.4 EXERCISES

1.Build a simulator for solving the example 1 through example 5 and Deutsch's difficult example to compare with the result of (Yoshimura and Kuh 1982) and that of (Chen and Liu 1984).

2. Test the result shown in this Chapter with your simulator.

3. Develop a neural network algorithm based on the HVH model.

4. Calculate the energy function E for the channel routing problems from Eq. (5. 4). Note that you may not consider the C-term.

5. The proposed algorithm introduced several heuristics in order to improve the frequency of the global minimum convergence: the modified McCulloch-Pitts neuron model, two forms of B-terms with modulo function, and the input saturations. In order to further improve it, think about new heuristics and try them.

Chapter 6

RNA SECONDARY STRUCTURE PREDICTION

In this Chapter two neural network models are used for solving RNA secondary structure prediction problems. One is for solving the maximum independent set problem and is slightly modified for predicting the RNA secondary structure. The other is modified from the graph planarization algorithm discussed in Chapter 4 and it takes advantage of molecular thermodynamics using a part of Tinoco's model. Both simulators were developed on a Macintosh machine, an HP Apollo 3500 workstation, and DEC 3100 workstations. A large number of simulation runs have been performed to test the proposed parallel algorithms and to observe the behavior of the proposed system. The simulation results demonstrated that the proposed parallel algorithms are promising and worthwhile for practical use. This Chapter is based on a paper published in IEEE Trans. on Neural Networks (Takefuji et al 1990b), a paper published in Biological Cybernetics (Takefuji, Lin, and Lee 1990a), and a paper published in Journal of Intelligent Manufacturing (Takefuji 1991).

6.1 INTRODUCTION

This Chapter first introduces a parallel algorithm for finding a near-maximum independent set of a circle graph within several hundred iteration steps. The proposed algorithm is modified for predicting the secondary structure of ribonucleic acids (RNA) where the circle graph is very suited for computing the secondary structure. Non-intersected edges in the circle graph provides information on the base pairs for the folding of the chains like the beads in a necklace. To generate the stable RNA structure, we want to maximize the number of non-intersected edges or base pairs. The proposed parallel algorithm is tested for predicting several RNA secondary structures.

Nucleic acids are the means by which information about the structure and function of a living organism is stored and passed on to the next generation. Nucleic acids is composed of only two types of molecules (nucleotides), deoxyribonucleic acids (DNA) and ribonucleic acids. Each nucleotide is itself made up of three smaller chemical units; a phosphate group, a sugar ring, and a nitrogenous base. The four different nucleotides of DNA have different bases but with identical phosphate and sugar groups: Adenine, Guanine, Cytosine, and Thymine as shown in Fig. 6-1 through Fig. 6-4.

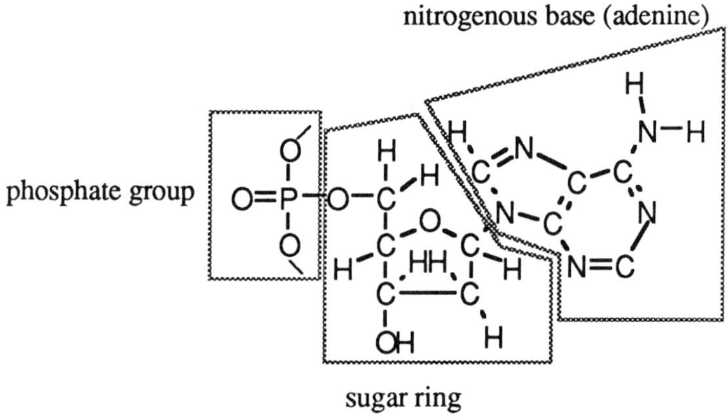

Fig. 6-1 Deoxyadenosine phosphate (A)-- a nucleotide

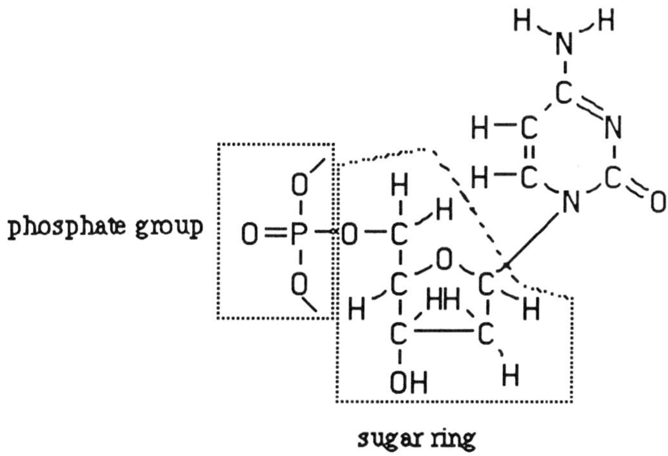

Fig. 6-2 Cytosine of DNA

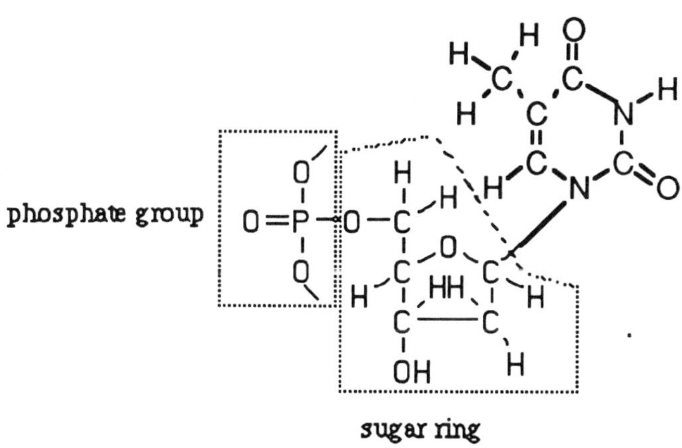

Fig. 6-3 Thymine of DNA

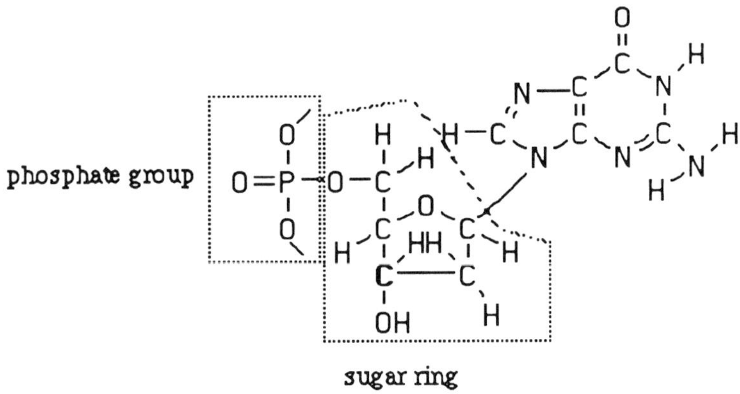

Fig. 6-4 Guanine of DNA

The organic bases of RNA include two compounds (Cytosine and Uracil) and two molecules (Adenine and Guanine). A base Uracil of RNA has a slightly different sugar ring from that of a base Thymine of DNA. RNA is identical to DNA except the sugar ring difference. When nucleotides form a stretch of DNA or RNA, the phosphate group of one nucleotide forms a chemical bond with the oxygen atom attached to the **bold** carbon atom of the next nucleotide as shown in Fig. 6-4. The double-helix forms when two separate DNA or RNA strands become linked together by weak chemical bonds between specific bases. The simple rules of this base-pairing are as follows: adenine always pairs with uracil while guanine pairs with cytosine. The primary structure is determined by the sequence of those bases (C, U, A, and G). The secondary structure is determined by the folding of the chains into a two-dimensional shape. The foldings of the chains into a three-dimensional shape is called the tertiary structure. Predicting the primary, the secondary and the tertiary structure is extremely important not only for fighting with viruses and genetic problems but also for enhancing the biotechnology. Tinoco's RNA structure stability model is used to compare our simulation result with the existing structures proposed by other investigators. The structure stability computation can show us how good our algorithm is.

The proposed parallel algorithm not only generates a near-maximum independent set of a circle graph but also predicts the secondary structure of ribonucleic acids. It requires n processing elements where n is the number of edges in the circle graph or the number of possible base pairs. Our simulator based on the proposed algorithm discovered the new structures in a sequence of 38 bases, a sequence of 55 bases, and a sequence of 359 bases from the potato spindle tuber viroid (PSTV), two of which are more stable than the formerly proposed structures. The simulator was tested by solving other problems.

We believe this is the first parallel/distributed processing attempt to solving RNA secondary prediction problems. This Chapter presents the clear comparison between the conventional RNA folding algorithms, the backpropagation algorithm by Quan and Sejnowski (Quan and Sejnowski 1989) or by Holley and Karplus (Holley and Karplus 1990), and our algorithm. Although the proposed algorithm is parallel computing, the simulator is currently running on sequential machines including a Macintosh machine, an HP Apollo 3500 computer and a DEC 3100 computer under Unix operating system. The state of the system can usually converge to the near-optimum solution within about 500 iteration steps. The algorithm uses n processing elements where each processing element performs the McCulloch-Pitts binary neuron.

6.2 MAXIMUM INDEPENDENT SET PROBLEMS

An independent set in a graph is a set of vertices, no two of which are adjacent. A maximum independent set is an independent set whose cardinality is the largest among all independent sets of a graph. The problem of finding a maximum independent set for arbitrary graphs is NP-complete (Karp 1972) (Garey and Johnson 1979). Gavril developed a $\theta(n^3)$ time algorithm for finding a near-maximum independent set in a circle graph where n is the number of edges in the circle graph (Gavril 1974). Supowit proposed an $O(n^2)$ time algorithm in the circle graph (Supowit 1987). Hsu gave an $O(m^4)$ time algorithm on planar perfect graphs where m is the number of vertices (Hsu 1988). Choukhmane (Choukhmane and Granco 1986) and Burns (Burns 1989) proposed an algorithm on cubic planar graphs. Masuda gave an O(n log n) time algorithm on circle graphs (Masuda and Nakajima 1988). Several parallel algorithms have been proposed by Karp (karp and Wigderson 1985), Luby (Luby 1986), and Goldberg (Goldberg and Spencer 1989) where the computation time is $O((\log m)^4)$.

The maximum independent set problem, the max cut problem, and the maximum clique problem are all interrelated each other for finding ground states of spin glasses with exterior magnetic field and solving circuit layout design problems in VLSI circuits and printed circuit boards.

Our parallel algorithm generates a near-maximum independent set of a circle graph in nearly constant time (Takefuji et al. 1990b). The algorithm uses n processing elements for an n-edge problem where each processing element performs the McCulloch-Pitts neuron model.

Consider the simple circle graph which has 14 vertices and 7 edges as shown in Fig. 6-5a. Fig. 6-5b shows the adjacency graph of Fig. 6-5a where the adjacency graph is generated by edge-intersection in the circle graph. For example the edge "d" intersects with three edges: c, e, and f. The adjacency graph G(V, E) of the circle graph in Fig. 6-5a is given by V={a, b, c, d, e, f, g} and E={(a b), (b c), (b e), (b f), (c d), (d e), (d f), (e f), (f g)}. A set of vertices {a, c, e, g} in Fig. 6-5c is the maximum independent set of this circle graph which is equivalent to the maximal planar subgraph as shown in Fig. 6-5d. In other words, to find the maximum independent set in the circle graph is equivalent to finding the maximum planar subgraph in the circle graph. In order to find the near-maximal planar subgraph in the circle graph with m-vertex and n-edge, n neurons (processing elements) are used in our algorithm. The output state of the ith neuron V_i=1 means that the *i*th edge is not embedded in the circle graph. The state of V_i=0 indicates that the *i*th edge is embedded in the circle graph.

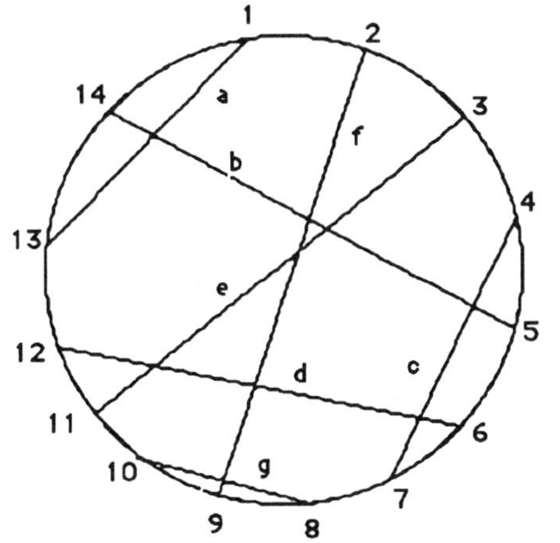

Fig. 6-5 (a) Circle graph with 14 vertices and 7 edges

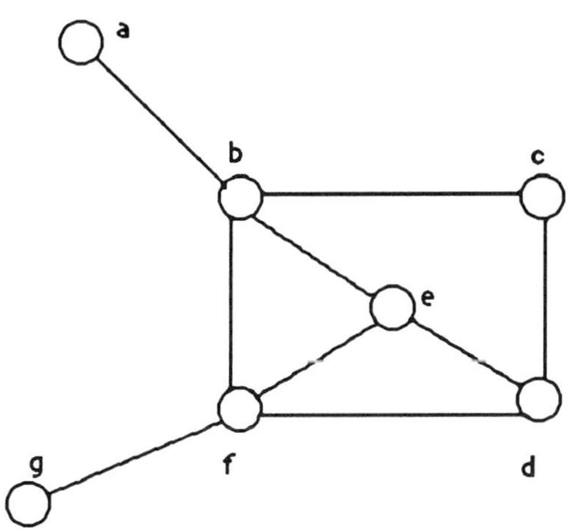

Fig. 6-5 (b) Adjacency graph of Fig. 6-5a

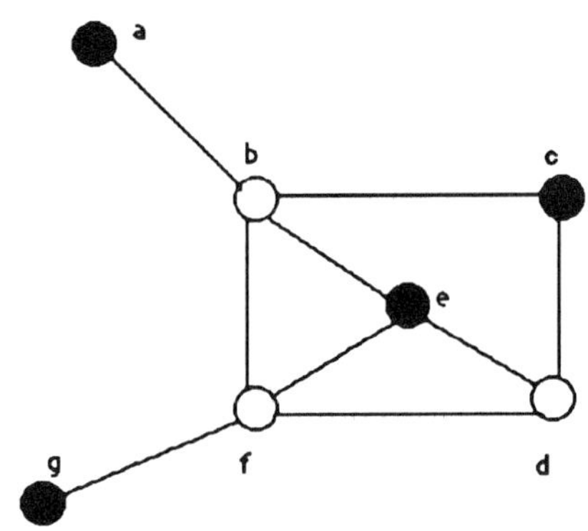

Fig. 6-5 (c) Maximum independent set of Fig. 6-5a

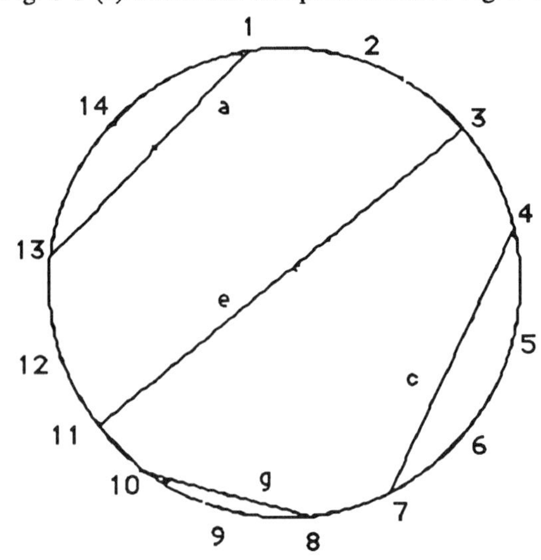

Fig. 6-5 (d) Maximum planar subgraph of Fig. 6-5a

The motion equation of the *i*th neuron for i=1,...,n is given by:

$$\frac{dU_i}{dt} = A\left(\frac{\sum_{j=1}^{n} d_{ij}(1-V_j)}{\text{distance}(i)}\right)(1-V_i) - Bh\left(\sum_{j=1}^{n} d_{ij}(1-V_j)\right)V_i \tag{6.1}$$

where $d_{xy}=1$ if the xth edge and the yth edge intersect each other in the circle graph, 0 otherwise. Note that A and B are constant coefficients. Edge-intersection conditions between the ith and the jth edges in the circle graph are given by: head(i)<head(j)<tail(i)<tail(j) and head(j)<head(i)<tail(j)<tail(i) where tail(i) and head(i) are two end vertices of the ith edge. Note that distance(i) is given by $\text{distance}(i)=\min(|\text{head}(i)-\text{tail}(i)|, |n+\text{head}(i)-\text{tail}(i)|)$ where tail(i)>head(i) is always satisfied. The function h(x) is 1 if x=0, 0 otherwise.

The first term is the inhibitory force in order to remove the edges which intersect with the ith edge in the circle graph. If the ith edge is removed from the circle graph, the first term will not be activated at all, because the state of the ith neuron should be $V_i=1$. In order to keep the ith edge in the circle graph, the first term should not have any edge-intersection violation. Whenever the ith edge have any edge-intersection violation, it will be eventually removed from the circle graph. The last term is the encouragement force to embed the ith edge in the circle graph. If the ith edge is removed but does not intersect with any other edges, the last term will force the ith neuron to be $V_i=0$. In other words, the ith edge is encouraged to exist in the circle graph. Fig. 6-6a shows the circle graph with 50 vertices and 122 edges. One of the solutions in Fig. 6-6a is shown in Fig. 6-6b.

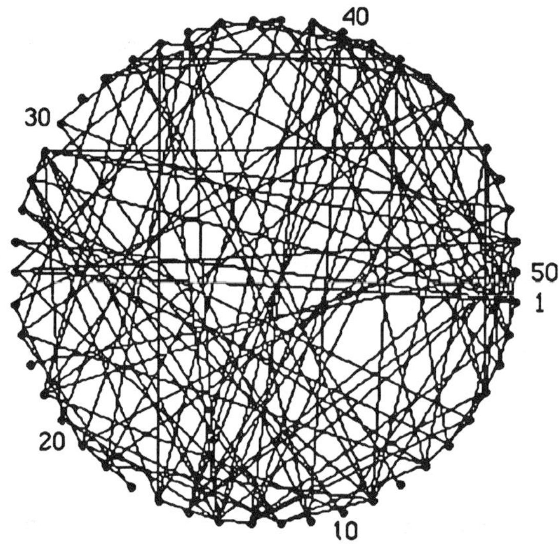

Fig. 6-6 (a) Circle graph with 50 vertices and 122 edges

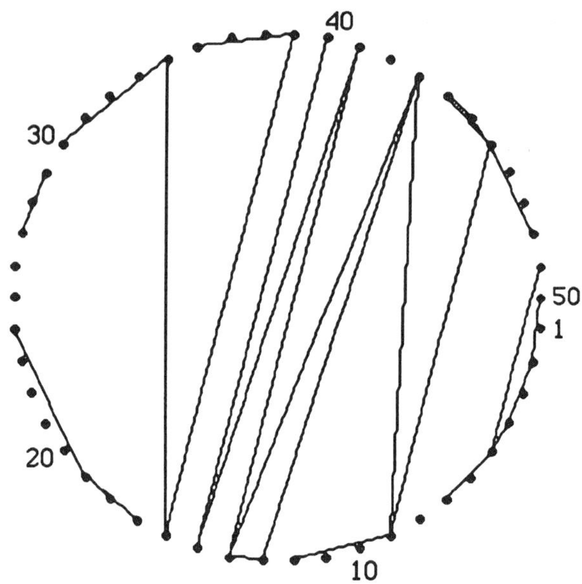

Fig. 6-6 (b) Near-maximum planar subgraph of the circle graph in Fig. 6-6a

Our simulator was modified for solving several secondary structure prediction problems in ribonucleic acids. In this Chapter only three examples are discussed. A sequence of 38 bases from residues 1118-1155 of E. coli 16S rRNA given by Stern (Stern et al. 1988), a sequence of 55 bases from R17 viral RNA (Tinoco et al. 1971), and a sequence of 359 bases from the potato spindle tuber viroid (PSTV) were used. Fig. 6-6a shows the secondary structure proposed by Stern (Stern et al. 1988) where the strength of structure stability is computed based on Tinoco's stability number (Tinoco et al. 1971): I. A-U pair, +1; II. G-C pair, +2; III. G-U pair, 0; IV. hairpin loops, -5 to -7;V. interior loops, -4 to -7; VI. bulges, -2 to -6. For the details of the Tinoco's stability model, see the paper (Tinoco et al. 1971). The stability number of the secondary structure in Fig. 6-7a is +7.

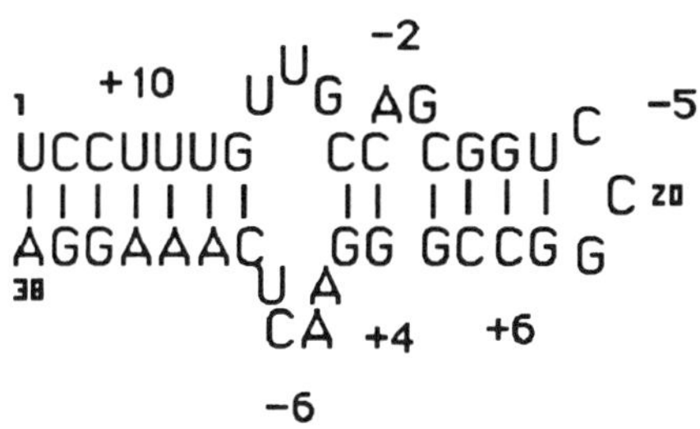

Fig. 6-7 (a) Secondary structure of a sequence of 38 bases proposed by Stern

6.3 PREDICTING THE RNA SECONDARY STRUCTURE

Fresco used the first RNA secondary structure model for predicting the secondary structure in Ribonucleic Acids (Fresco et al. 1960). Two types of RNA folding algorithms have been reported: the "combinatorial" method introduced by Pipas (Pipas and McMahon 1975) and the "recursive" or dynamic programming method introduced by Nussinov (Nussinov et al. 1978). Both algorithms including the latest method proposed by Zuker (Zuker 1989) are all based on the sequential computation. Unfortunately few parallel algorithms based on molecular thermodynamics models have been reported. Recently Qian and Sejnowski (Qian and Sejnowski 1988), and

Holley and Karplus (Holley and Karplus 1989) have reported a backpropagation algorithm using a three-layer-feed-forward neural network for protein secondary structure prediction. Their method is based on the correlation between secondary structure and amino acid sequences. However they have the following drawbacks over the conventional RNA folding algorithms based on molecular thermodynamics models: **1**. They need a teacher to force the network to learn the correlation between secondary structure and amino acid sequences. **2**. The correlation models cannot provide accurate prediction if a completely uncorrelated new datum is given where the previously learned correlation information is useless. **3**. Their feed-forward neural network requires a prohibitively long learning process to deal with a long sequence of bases for the RNA secondary structure prediction. Their methods cannot be applied to large-size prediction problems unless prohibitively long learning time is permitted. **4**. No theorem is given to determine the neural network architecture including how many hidden layers and how many hidden neurons per hidden layer should be used.

Our algorithm requires neither a teacher nor a learning process. The proposed parallel algorithm using n processors (where n is the number of possible base pairs) can yield the suboptimum solution within several hundred iteration steps (Takefuji et al. 1990b). Our goal is to maximize the number of edges in the planar circle graph where an edge represents a possible base-pair such as G-C base pair or A-U base pair. Diener supports our goal of maximizing the number of base pairs for predicting the secondary structure in viroids (Diener 1987).

The motion equation of Eq. (6.1) is slightly modified and used for predicting the secondary structure of Ribonucleic Acids. One is that edge-intersection violation conditions must be updated. The total of six conditions to describe the edge-intersection in the circle graph are required: head(i)<head(j)<tail(i)<tail(j), head(j)<head(i)<tail(j)<tail(i), tail(i)=tail(j), tail(i)=head(j), head(i)=head(j), and head(i)=tail(j). The last four violation conditions are newly added to the first two conditions, because a single base must not involve in more than one base-pair. The other modification is in the distance(i) function of Eq. (6.1) where distance(i) is given by distance(i)=|head(i) -tail(i)|.

A sequence of m bases is given to the simulator. It generates the circle graph with m vertices and n edges where n is the number of possible base pairs including

G-C base pairs and A-U base pairs. The possible base-pair must also satisfy the hairpin-loop constraint such as |head(i) -tail(i)|>3. Because Tinoco (Tinoco et al. 1971) stated that it is sterically impossible to organize the hairpin loop with less than three bases. The circle graph is fed to the neural network simulator in order to find the near-maximum independent set.

Our simulator was tested by solving several secondary structure prediction problems in ribonucleic acids. In this Chapter three examples are only shown. A sequence of 38 bases from residues 1118-1155 of E. coli 16S rRNA given by Stern (Stern et al. 1988) was used. Fig. 6-7a shows the secondary structure proposed by Stern where the strength of structure stability is computed based on Tinoco's stability number (Tinoco et al. 1971). The stability number of the secondary structure in Fig. 6-7a is +7. Fig. 6-7b shows the circle graph with 38 vertices and 151 edges. Remember that each edge represents a possible base-pairing.

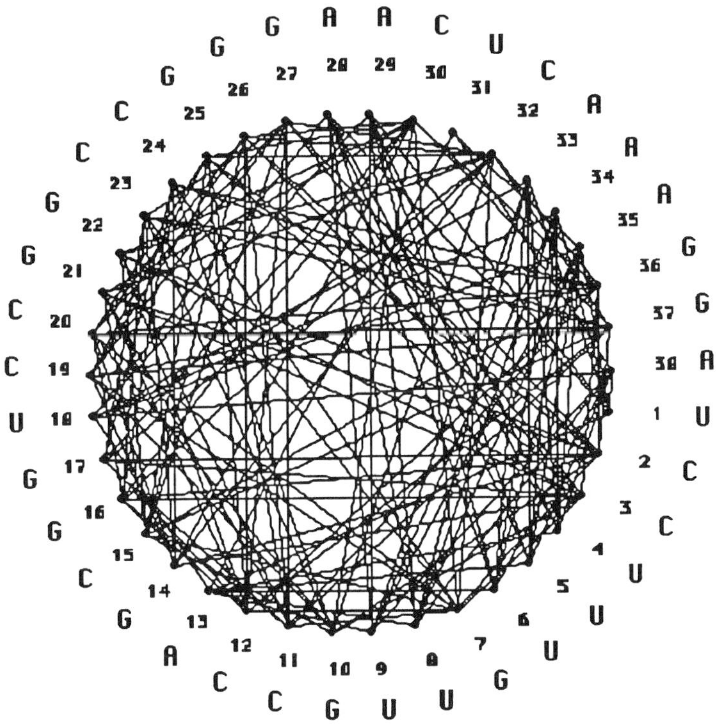

Fig. 6-7 (b) Circle graph with 38 vertices and 151 edges

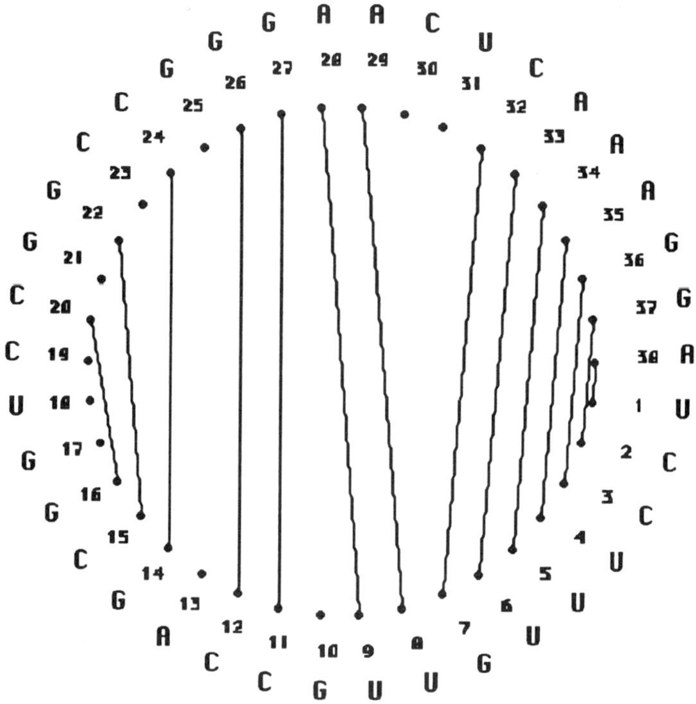

Fig. 6-7 (c) State of the system after the 104 iteration step

When A=B=1 and $U_i(0)$=-5 for i=1,..., 151, the state of the system converged to the solution in the 104th iteration step. Fig. 6-7c shows the simulation result which contains 14 edges. The secondary structure of the simulation result is given in Fig. 6-7d.

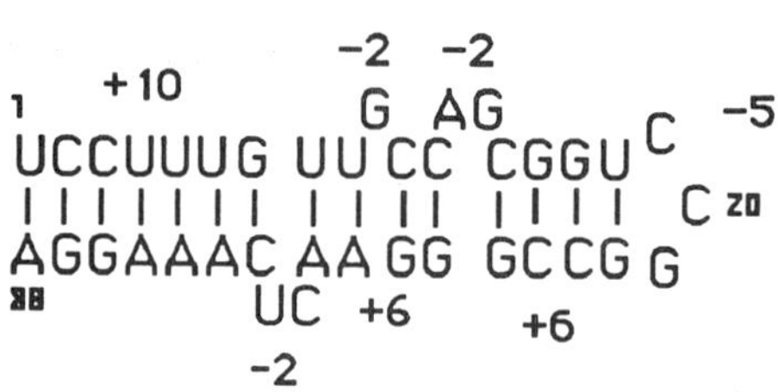

Fig. 6-7 (d) Secondary structure predicted by our algorithm

The stability number of the structure is +11. It indicates that the simulator found more stable structure than that of Stern's (Stern et al. 1988).

Secondly a sequence of 55 bases from R17 viral RNA (Tinoco et al. 1971) was investigated. Fig. 6-8a shows the circle graph with 55 vertices and 331 edges. When the following parameters were used; A=1, B=0.01, and $U_i(0)$=-5 for i=1,..., 331, the state of the system converged to the solution in the 161th iteration step. The simulation result is shown in Fig. 6-8b. Our algorithm embedded 20 edges in the circle graph where the structure stability is +7, while the stability of the secondary structure proposed by Tinoco as shown in Fig. 6-8c is +8. Fig. 6-8d shows the secondary structure of the simulation result in Fig. 6-8b.

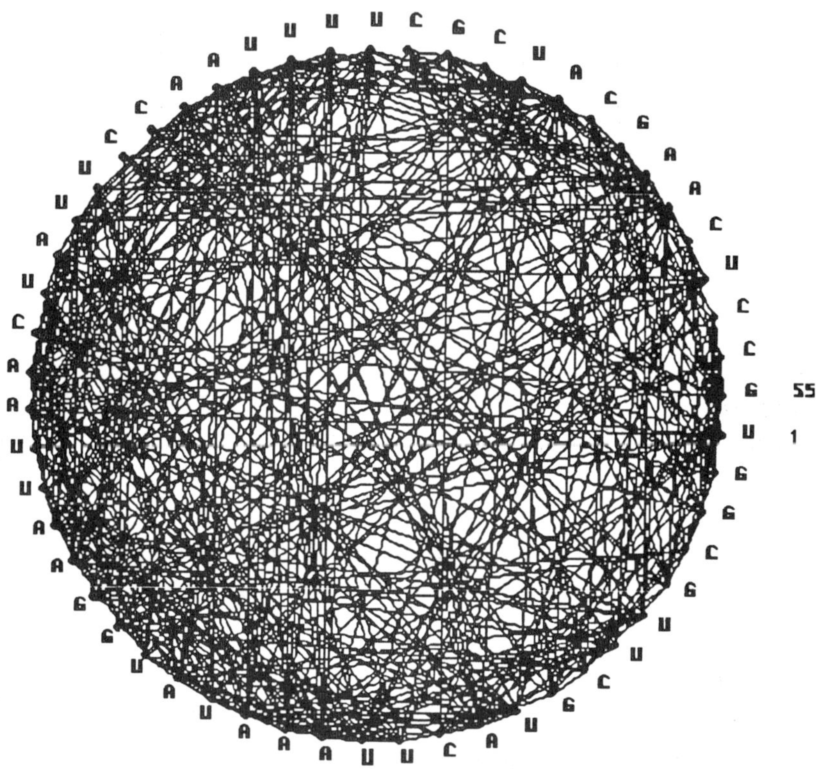

Fig. 6-8 (a) Circle graph with 55 vertices and 331 edges

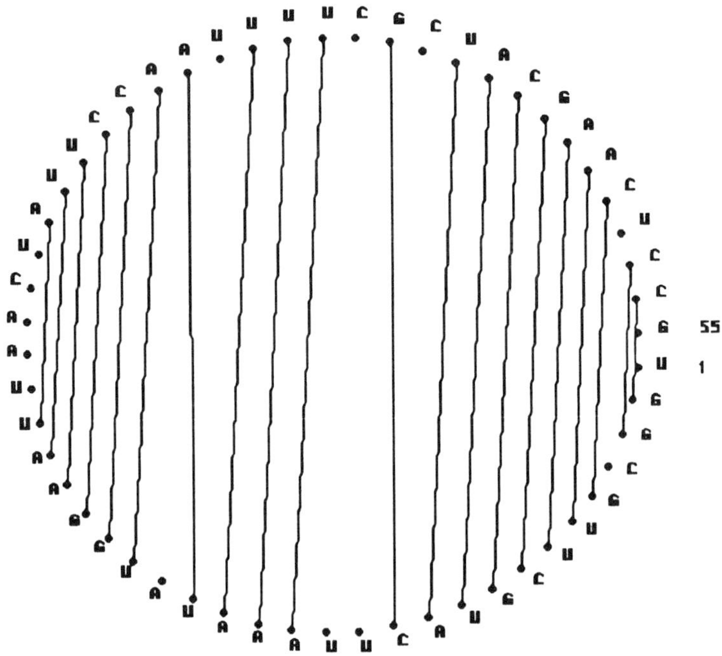

Fig. 6-8 (b) State of the system after the 161th iteration step

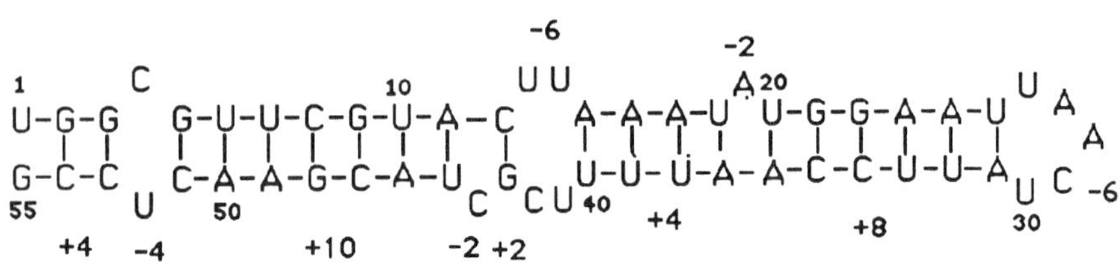

Fig. 6-8 (c) Secondary structure predicted by Tinoco

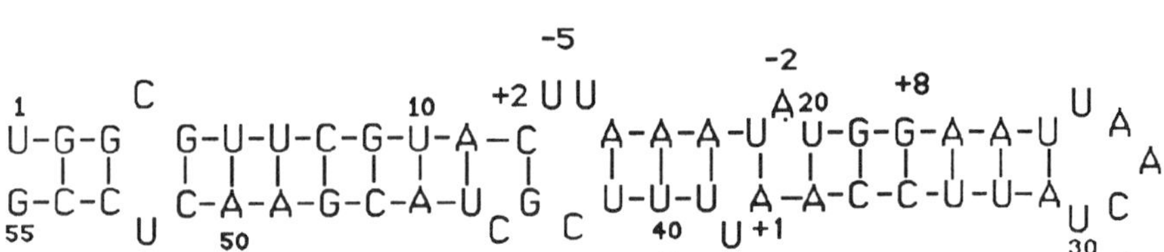

Fig. 6-8 (d) Secondary structure predicted by our algorithm

Finally a sequence of 359 bases from the potato spindle tuber viroid (PSTV) was used to verify our algorithm. Gross (Gross et al. 1978) proposed the secondary structure of the PSTV where the stability number is +62. The circle graph with 359 vertices and 1017 edges was generated where possible base pairs (i and j) were given by the following condition: $350<i+j<370$. The state of the system converged to the solution in the 240th iteration step with A=1, B=0.01, and $U_i(0)=-5$ for i=1,..., 1017. The simulation result was obtained when A=1, B=0.01, and small negative random numbers are assigned to $U_i(0)$ for i=1,..., 1017. The secondary structure predicted by our algorithm is composed of 359 vertices and 128 edges where the stability number is +65. Sanger proposed another secondary structure of the PSTV where the stability number is +64 (Sanger 1984). It indicates that our simulator found the most stable structure of the PSTV. Our simulation result shows that within about 500 iteration steps the state of the system can converge to the solution in the PSTV secondary structure prediction problem. The details of the PSTV experiment are depicted in (Takefuji 1991).

In this Section we have shown the parallel algorithm for the maximum independent set problem which is modified for predicting the secondary structure of ribonucleic acids. The algorithm uses n processing elements where n is the number of edges in the circle graph or the number of possible base-pair in ribonucleic acids. Our simulation result shows that the state of the system converges to the solution within several hundred iteration steps. The simulator discovered the most stable structures in a sequence of 38 bases and a sequence of 359 bases from the PSTV within 500 iterations.

6.4 PLANARIZATION AND RNA STRUCTURE PREDICTION

The stability number for a given RNA secondary structure is the sum of the contributions of the loops, bulges, and helices. The structure with the highest number is the most stable, called optimal folding. The mathematical problem to compute an optimal folding based on free-energy minimization is mapped onto a graph planarization problem which was discussed in Chapter 4. In the planarization problem we want to maximize the number of edges in a plane with no two edges crossing each other. An A-U or G-C base pairs are only considered as possible edges to be embedded in a plane while the bases are the vertices. In other words, for a fragment stretching from ribonucleotides i to j, it is denoted by the subscript ijth neuron where the output and the input is depicted by V_{ij} and U_{ij} respectively for $i=1,...,n-1$ and $j=i+1,...,n$.

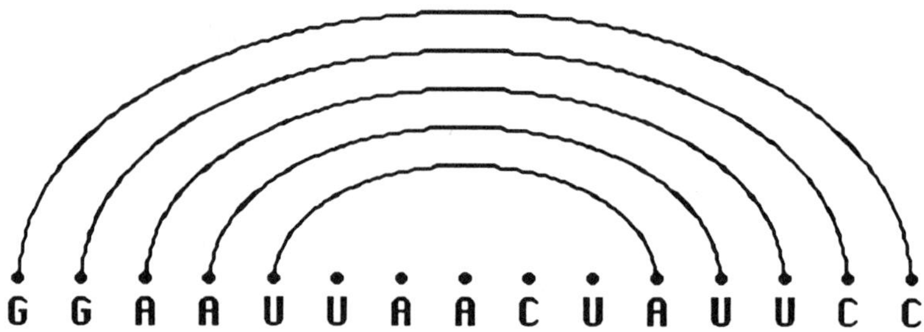

Fig. 6-9 (a) Single-row representation of fifteen bases

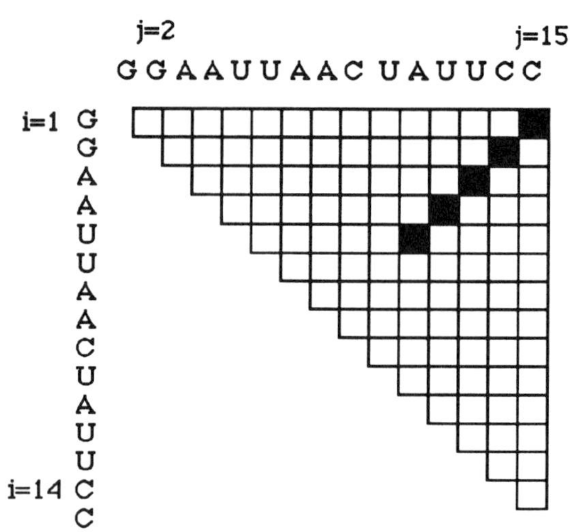

Fig. 6-9 (b) Neural representation of a sequence of fifteen bases

Consider a sequence of fifteen bases as shown in Fig. 6-9a. In our algorithm a single-row representation is used where five edges are embedded. A (15 x 14)/2 neural network array in Fig. 6-9b is used to predict the secondary structure of this problem. The following five functions are considered: b(i,j), g(i,j,k), f(i,j), p(i,j,t), and h(x). The function b(i,j) denotes the possible pairing: b(i,j)=1 if i and j bases are one of four base pairs (G-C, C-G, A-U, or U-A base pairs), 0 otherwise. The cross bonding is sterically impossible so that the graph must be planar where two edges must not cross each other. A violation function g(i,j,k) for graph planarization was described in our paper (Takefuji and Lee 1989): g(i,j,k)=1 if i<j<k, 0 otherwise. The function f(k,l) indicates the strength of a base pair bond between k and l bases: f(k,l)=2 if k and l bases are a G-C pair, 1 if they are an A-U pair, 0 otherwise. This strength of a base pair bond is based on the result of Tinoco (Tinoco et al. 1971). The hairpin loop constraint is also considered in our algorithm where more than two bases are required to make a hairpin loop. The hairpin constraint function p(i,j,t) is given by: p(i,j,t)=1 if (j-hairpin)<i, 0 otherwise where hairpin is given by hairpin=4 if t=0, 55-t if 55-t >4, 4 otherwise. h(x) is the hill-climbing function, 1 if x=0, 0 otherwise.

To predict suboptimal foldings of a sequence of n bases, n(n-1)/2 neurons are required. The motion equation of the ijth neuron is given by:

$$\begin{aligned}\frac{dU_{ij}}{dt} &= -A_1\left(\sum_{k\neq i} V_{ik}-1\right)b(i,j)-A_2\left(\sum_{k\neq j} V_{jk}-1\right)b(i,j)\\ &-B_1 \sum_{k<i<l<j} V_{kl}g(k,i,l)g(i,l,j)f(k,l)\\ &-B_2 \sum_{i<k<j<l} V_{kl}g(i,k,j)g(k,j,l)f(k,l) -Cp(i,j,t) +Dh\left(\sum_{k\neq i} V_{ik}\right)\end{aligned} \quad (6.2)$$

where V_{xy} is V_{xy} if x<y, V_{yx} otherwise. The first term in Eq. (6.2) forces the *i*th base to have one and only one bond. Note that if the ith base has strong violations caused by other bases, it cannot have any bond. The second term also forces the jth base to have one and only one bond. The third and fourth terms are always inhibitory which always satisfy planarization conditions. The fifth term is the inhibitory hairpin constraint which prohibits less than three bases to make a hairpin loop. The last term is a hill-climbing force which allows the state of the system to escape from the local

minimum.

Consider a sequence of 55 bases from R17 viral RNA (Tinoco et al. 1971). 1485 (=55X54/2) neurons are used to solve this problem. Fig. 6-10a and 6-10b show the state of the system after the first iteration and the sixty-first iteration respectively. Fig. 6-11 shows the predicted secondary structure of a sequence of 55 bases as shown in Fig. 6-10b. The total stability of the structure is +7 which is equivalent to the free energy ΔG=-8.4kcal. When pairs (A^{15} to U^{41}), (A^{16} to U^{40}), and (A^{17} to U^{39}) are shifted to (A^{15} to U^{40}), (A^{16} to U^{39}), and (A^{17} to U^{38}) respectively, the total stability of the structure becomes +8 which is the optimum structure (Tinoco et al. 1971).

The proposed algorithms for predicting RNA secondary structures play an important role in molecular biology and manufacturing.

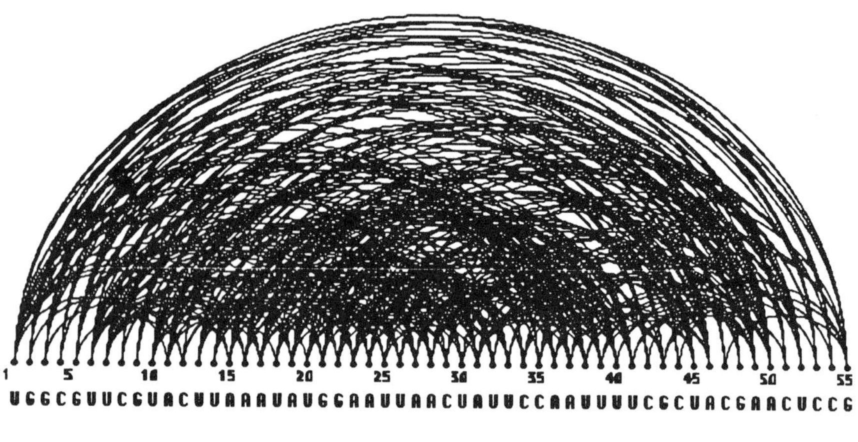

Fig. 6-10 (a) State of the system after the first iteration step

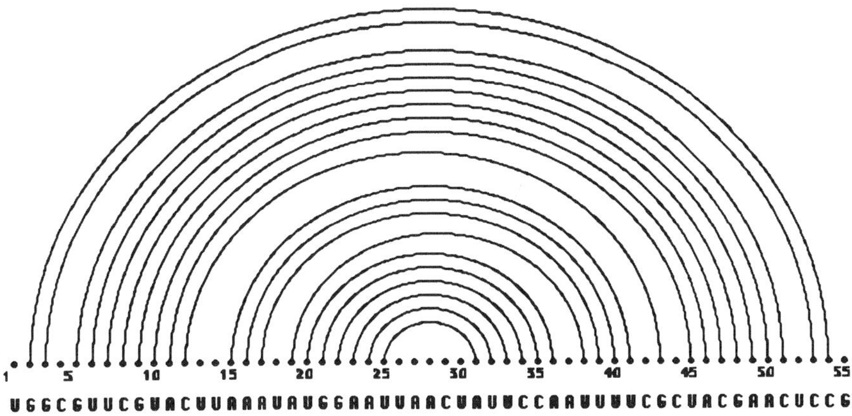

Fig. 6-10 (b) State of the system after the sixty-first iteration step

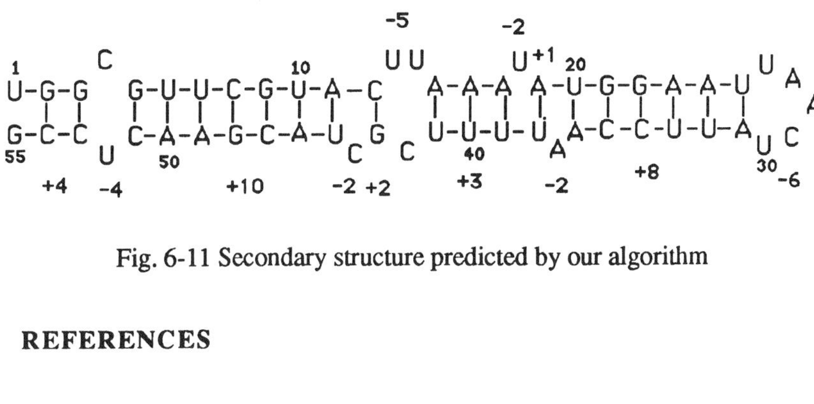

Fig. 6-11 Secondary structure predicted by our algorithm

6.5 REFERENCES

Burns J. E., 1989, The maximum independent set problem for cubic planar graph. Networks, 19.

Choukhmane E., and Franco J., 1986, An approximation algorithm for the maximum independent set problems in cubic planar graphs. Networks, 16.

Diener T. O., 1987, The Viroid (Plenum Press, New York).

Fresco J. R., Alberts B. M., and Doty P., 1960, Some molecular details of the secondary structure of ribonucleic acid. Nature, 188, 98.

Garey M. R., and Johnson D. S., 1979, Computers and Intractability: A Guide to the theory of NP-completeness (W. H. Freeman and Company).

Gavril F., 1974, Algorithms on circular-arc graph. Networks, 4, 357-369.

Goldberg M., and Spencer T., 1989, A new parallel algorithm for the maximum independent set problem. SIAM J. Comput., 18, 2.

Gross H. J. et al., 1978, Nucleotide sequence and secondary structure of potato sprindle tuber viroid. Nature, 273.

Holley L. H., and Karplus M., 1989, Protein secondary structure prediction with a neural network. Proc. Natl. Acad. Sci. USA, 86.

Hsu W. L., 1988, The coloring and maximum independent set problems on planar perfect graphs. J. of ACM, 35, 3.

Karp R. M., 1972, Complexity of Computer Computations (R. E. Miller and J. W. Thatcher, Eds, Plenum Press, New York).

Karp R. M., and Wigderson A., 1985, A fast parallel algorithm for the maximum independent set problem. J. of ACM, 32, 4.

Luby M., 1986, A simple parallel algorithm for the maximum independent set problem. SIAM J. Comput., 15, 4.

Masuda S., and Nakajima K., 1988, An optimal algorithm for finding a maximum independent set of a circular-arc graph. SIAM J. Comput., 17, 1.

McCulloch W. S., and Pitts W. H., 1943, A logical calculus of ideas immanent in nervous activity. Bulletin of Mathematical Biophysics, 5, 115.

Nussinov R., Pieczenik G., Griggs J. R., and Kleitman D. J., 1978, Algorithm for loop matching. SIAM J. Appl. Math., 35, 68.

Pipas J. M., and McMahon J. E., 1975, Method for predicting RNA secondary structure. Proc. Natl. Acad. Sci. U.S.A.,72, 2017.

Qian N., and Sejnowski T., 1988, Predicting the secondary structure of globular proteins using neural network models. J. Mol. Biol., 202, 865.

Sanger H. L., 1984, Minimal infectious agents, (B. W. J. Mahy and J. R. Pattison, eds., Stern S., Weiser B. , and Noller H. F., 1988, Model for the three-dimensional folding of 16 S ribosomal RNA. J. Mol. Bio., 204.

Supowit K. J., 1987, Finding a maximum planar subset of a set of nets in a channel. IEEE Trans. CAD, 6, 1.

Takefuji Y., and Lee K. C., 1989, A near-optimum parallel planarization algorithm. Science, 245, 1221-1223.

Takefuji Y., Lin C. W., and Lee K. C., 1990a, A parallel algorithm for estimating the secondary structure in Ribonucleic Acids. Biological Cybernetics, 63, 337-340.

Takefuji Y., Chen L. L., Lee K. C., and Huffman J., 1990b, Parallel algorithms for finding a near-maximum independent set of a circle graph. IEEE Trans. on Neural Networks, 1, 3, 263-267.

Takefuji Y., 1991, Neural network parallel computing for combinatorial optimization problems, Journal of Intelligent Manufacturing.

Tinoco I., Uhlenbeck O. C., and Levine M .D, 1971, Estimation of secondary structure in ribonucleic acids. Nature, 230.

Zuker M., 1989, On finding all suboptimal foldings of an RNA molecule. Science, 244, 48.

Cambridge University Press, London).

6.6 EXERCISES

1. Build a simulator based on Eq. (6.1) and observe the behavior of the system in terms of the solution quality and the number of iteration steps. Compare with the best known algorithm.
2. Build a simulator based on Eq. (6.2) and test the system.
3. Discuss the advantages and the disadvantages of two algorithms.

Chapter 7

KNIGHT'S TOUR PROBLEMS

7.1 INTRODUCTION

The earliest serious attempt to find a knight's tour on the chessboard was made by Leonhard Euler in 1759 (Euler 1766). A knight must traverse all of the squares on an m x n chessboard but visit every square once and only once and return to the originated square where the knight moves in an L-shape route. Many great mathematicians including De Moivre, Vandermonde (Vandermonde 1774), Warnsdorff (Warnsdorff 1823), Pratt (Pratt 1825), Roget, Legendre (Legendre 1830), and De Lavcrncdc (Lavernede 1839) attempted to solve the general problem. Their methods are based on either divide-and-conquer or backtracking dedicated for solving the 8 x 8 chessboard knight's tour problem. No general methods have been given to the problem in the last two centuries (Ball and Coxeter 1974). W. W. R. Ball and H. S. M. Coxeter reintroduced the problem in their book in 1892 and 1974. A parallel algorithm for finding a knight's tour on the m x n chessboard is presented in this paper. The problem is to find a constrained Hamiltonian circuit which belongs to NP-complete problems. However we are not sure whether the knight's tour problem is NP-complete or not. The artificial neural network computing makes it possible to provide the elegant algorithm in parallel which takes advantage of the simplied biological neural computation. Several examples are shown to demonstrate the capability of our parallel algorithm. This Chapter is based on a paper published in IEEE Trans. on System, Man, and Cybernetics (Lee and Takefuji 1992).

7.2 NEURAL REPRESENTATION AND MOTION EQUATIONS

In this Chapter a two-dimensional triangular neural network representation is introduced. The system requires p(p-1)/2 processing elements or hysteresis McCulloch-Pitts neurons where p is the number of squares on an m x n chessboard. Each state of the neurons represents a path. In order to establish a knight's path on a square on the chessboard, two neurons must be fired because one is for coming into the square and the other is for going out from the square. In our algorithm the following motion equation for the *ij*th neuron is used to solve the problem:

if $d_{ij} = 1$ then

$$\frac{dU_{ij}}{dt} = -\left(\sum_{k=1}^{p} V_{ik}d_{ik} - 2\right) - \left(\sum_{k=1}^{p} V_{kj}d_{kj} - 2\right)$$

if $d_{ij} = 0$ then

$$\frac{dU_{ij}}{dt} = 0 \qquad (7.1)$$

where the upper triangular elements in the two-dimensional array are only used: V_{ij} for i<j and V_{ij} for i>j is given by V_{ji}. The state of V_{ij} actually represents a knight's tour between the *i*th and the *j*th square. In other words, the upper triangular neural array represents the nondirected adjacency matrix in order to find a Hamiltonian circuit. In Eq. (7.1) d_{ij} is 1 if the move from the *i*th square to the *j*th square is legal or valid, and 0 otherwise. The first term and the second term are to determine a move between the *i*th and the *j*th square if it is valid. The first term forces the *ij*th neuron to have two valid moves from the *i*th square and the second term from the *j*th square respectively. The state of V_{ij} is updated by the hysteresis McCulloch-Pitts function: $V_{ij}=f(U_{ij})=1$ if $U_{ij}>3$, 0 if $U_{ij}<0$, and no changes otherwise.

Fig. 7-1 shows where the knight moves in an L-shape route. In a 3 x 4 chessboard knight's tour problem, 12 squares are numbered from left to right and from top to bottom as shown in Fig. 7-2a. From #2 square there are three valid moves: $V_{2,8}$, $V_{2,9}$ and $V_{2,11}$ where two moves are only needed to find the Hamiltonian circuit.

Fig. 7-2b shows the p(p-1)/2=66 neurons for this problem where the black squares indicate their outputs are one's. Fig. 7-2b depicts the two closed loops as shown in Fig. 7-2c which is not the Hamiltonian circuit. Unfortunately there is no solution for the 3 x 4 knight's tour problem (17). In 1943 Fred. Schuh stated that the number of squares has to be even which is necessary but not sufficient (17).

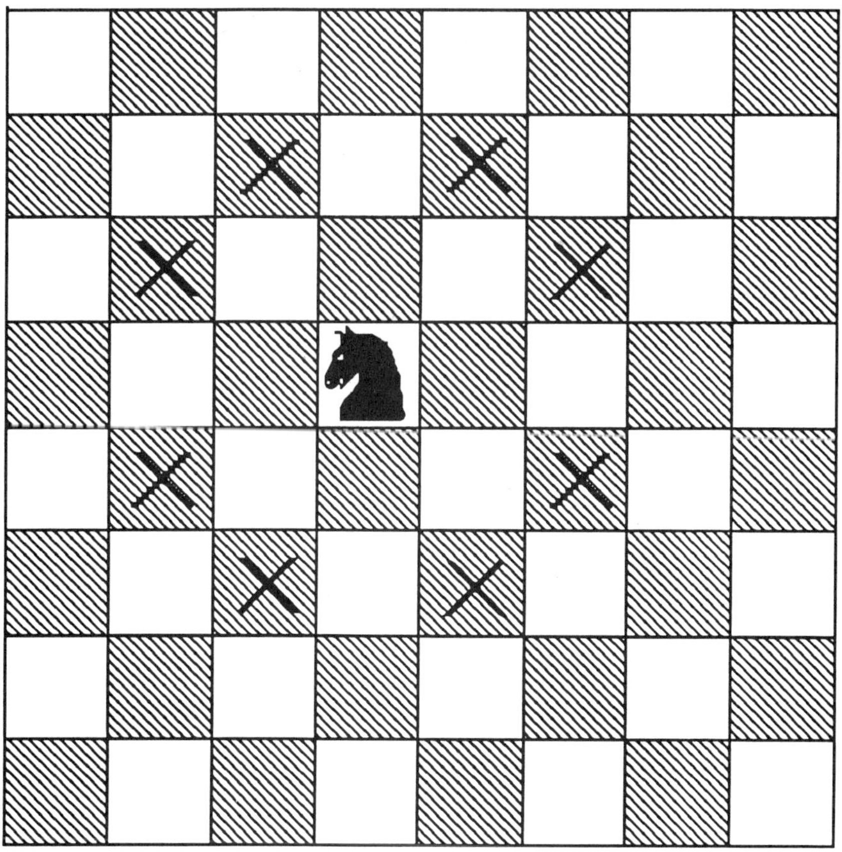

Fig. 7-1 (a) The knight moves in an L-shape route

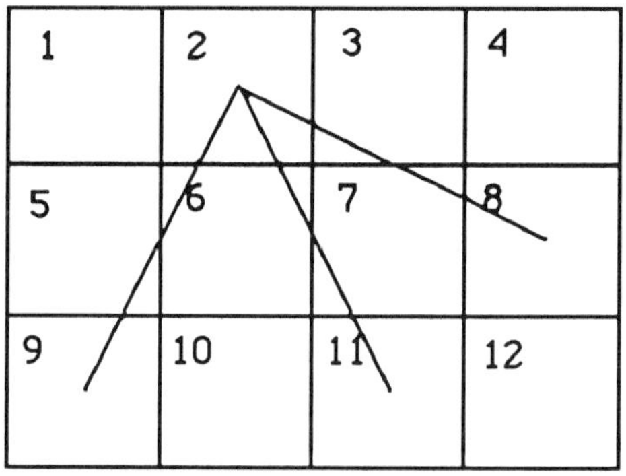

Fig. 7-2 (a) a 3 x 4 chessboard

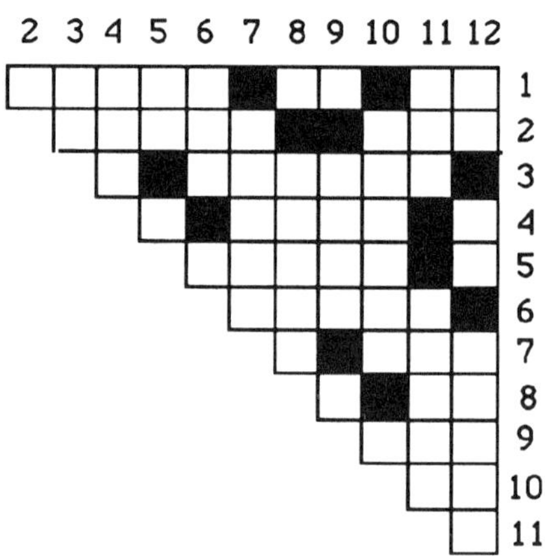

Fig. 7-2 (b) Neural representation

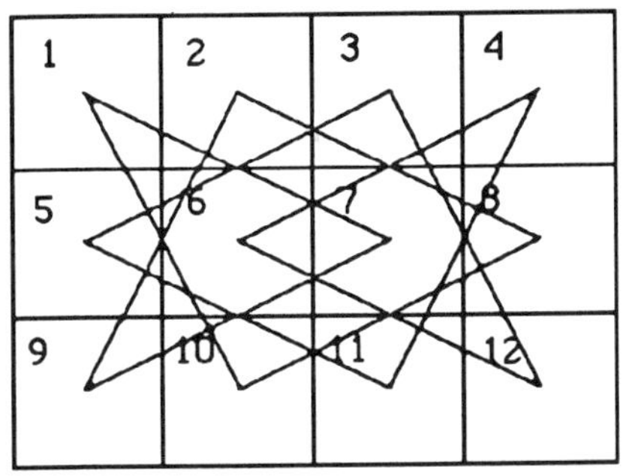

Fig. 7-2 (c) unsatisfactory solution

We have developed the simulator on a Macintosh SE/30 based on Eq. (7.1) to verify our algorithm. Fig. 7-3 through 7-10 show solutions for 6 x 6, 8 x 8, 10 x 10, 12 x 12, 14 x 14, 16 x 16, 18 x 18, and 20 x 20 chessboard knight's tour problems respectively.

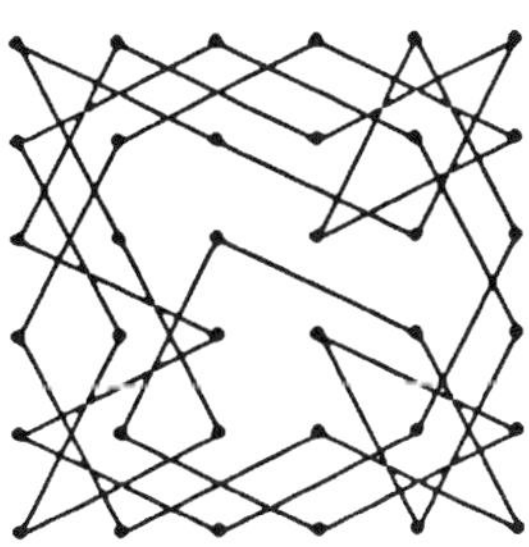

Fig. 7-3 6 x 6 chessboard

Fig. 7-4 8 x 8 chessboard

The average number of iteration steps takes less than 100 steps. We have observed that the larger problem the more often the state of the system converged to the unsatisfactory solution with several closed loops. Without the hysteresis property, the state of the system tends to oscillate and hardly converges to the solution.

Fig. 7-5 10 x 10 chessboard

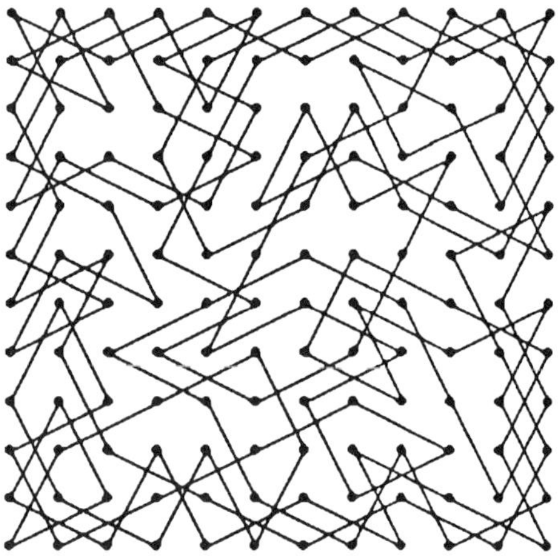

Fig. 7-6 12 x 12 chessboard

7.3 REFERENCES

Ball R.W. W. and Coxeter H. S. M., Mathematical recreations & essays, 1892 for 1st edition and 1974 for 12th edition, University of Toronto Press.

Euler L., (1766), Memoire de Berlin for 1759, Berlin, pp.310-337.

Lavernede De, (1839), Memoires de l'Academie royale du Gard, Nimes, pp.151-179.

Lee K. C. and Takefuji Y., (1990), Finding a knight's tour on an N x M chessboard with O(NM) hysteresis McCulloch-Pitts neurons, submitted to IEEE Trans. on System, Man, and Cybernetics, 1990.

Legendre, (1830), Theorie des Nombres, Paris 2nd edition, 2, p. 165.

McCulloch W. S. and Pitts W. H., (1943), Bulletin of Mathematical Biophysics, 5, 115.

Pratt, (1825), Studies of Chess, sixth edition, London.

Schuh Fred., (1968), the master book of mathematical recreations, (Dover Pub.).

Vandermonde, (1774), L'Historie de l'Academie des Sciences for 1771, Paris, pp.566-574.

Warnsdorff H. C., (1823), Des Rosselsprunges einfachste und allgemeinste Losung, Schmalkalden.

7.4 EXERCISES

1. Build a simulator to verify Eq. (7.1) and solve several knight's tour problems.

2. Discuss the minimum number of required neurons. Give the minimum number of neurons for the knight's tour problem on the 8 x 8 chessboard.

Fig. 7-7 14 x 14 chessboard

Fig. 7-8 16 x 16 chessboard

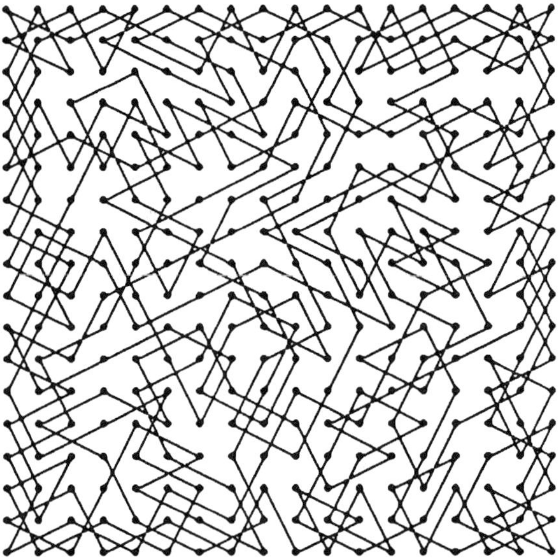

Fig. 7-9 18 x 18 chessboard

Fig. 7-10 20 x 20 chessboard

Chapter 8

SPARE ALLOCATION PROBLEMS

A parallel algorithm for spare allocation problems is introduced in this Chapter. In manufacturing memory chips, the Redundant Random Access Memory (RRAM) technology has been widely used because it not only provides the repairability of faulty cells but also enhances the production yield. RRAM has several rows and columns of spare memory cells which are used to replace the faulty cells. The goal of our algorithm is to find a spare allocation which repairs all the faulty cells in the given faulty cell map. The proposed parallel algorithm requires 2n processing elements for the $n \times n$ faulty cell map problem. The algorithm is verified by a large number of simulation runs. Under the simulation the algorithm finds one of the near-optimum solutions in a nearly constant time with O(n) processors. This Chapter is based on a paper published in IEEE Trans. on Reliability (Funabiki and Takefuji 1991).

8.1 INTRODUCTION

With the advancement of the VLSI technology, an integrated system interconnecting massive identical elements such as memory cells, CCD cells, or processing elements can be built on a silicon chip or wafer. However, as the density of the VLSI system increases, the probability of defects in the system also increases, and the production yield tends to decrease. In manufacturing memory chips, the Redundant Random Access Memory (RRAM) technology has been widely used because it not only provides the repairability of faulty cells but also enhances the production yield. RRAM has several rows and columns of spare memory cells. When a faulty cell is detected in a device test, the faulty row or column which contains the faulty cells is disconnected and is replaced by a spare per row or spare per column (Schuster S. E. 1978), (Cenker R. P. et al. 1979), (Fitzgerald B. F. and Thoma E. P. 1980), (Stapper C. H., McLaren A. N. , and Dreckmann M., 1980), (Sud R. and Hardee K. C. 1980), (Evans R. C. 1981), (Hayasaka Y. , Shimotori K., and Okada K. 1982), (Benevit C. A. et al. 1982), (Carlson R. O. and Neugebauer C. A. 1986). This redundancy technology can be also used in manufacturing other integrated systems.

A pattern of faulty cells in an $n \times n$ matrix cells is described by an $n \times n$ fault map F. In the fault map F, each element f_{ij} represents the condition of the corresponding cell, where $f_{ij}=1$ means that the cell at (i,j) has defects and $f_{ij}=0$ means that the cell has no defect. Fig. 8-1 shows an 8×8 fault map problem where two spares per row and two spares per column are given. In the fault map, the black squares indicate faulty cells and the white squares indicate normal cells. The fault map contains seven faulty cells at (1,3), (1,7), (3,5), (4,1), (4,8), (7,5), and (8,2). These faulty cells must be repaired by spares per row or spares per column, or both of them.

The goal of our algorithm is to find a spare allocation where all the faulty cells in the given fault map are repaired or covered by the given spares per row and spares per column. Several polynomial

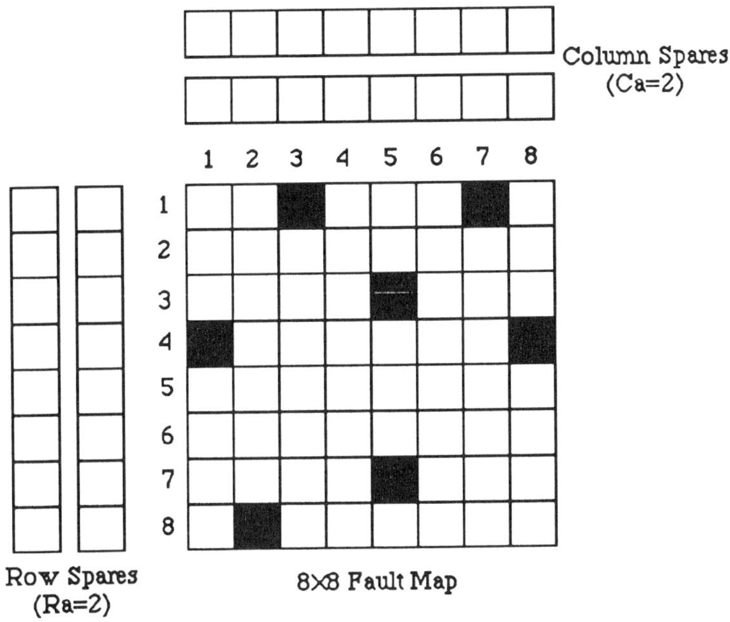

Fig. 8-1 8×8 fault map problem with two spares per row and two spares per column

time sequential algorithms for this problem have been proposed in RRAM applications (Tarr M., Boudreau D., and Murphy R. 1984), (Day J. R. 1985), (Kuo S-Y. and Fuchs W. K. 1987), (Wey C-L. and Lombardi F. 1987), (Haddad R. W. and A. T. Dahbura A. T. 1987), (Huang W. K., Shen Y-N., and Lombardi F. 1990). In 1984, Tarr et al. proposed two algorithms ; the broadside approach and the repair-most approach (Tarr M., Boudreau D., and Murphy R. 1984). The broadside approach is to scan the memory and to replace each faulty cell with a spare memory per row or a spare memory per column whichever is available where no optimization is performed. The repair-most approach is to repeatedly repair the row or column which has the most faulty cells. In 1985, Day proposed the fault-driven approach which is to repair the faulty cells according to user-defined preferences (Day J. R. 1985). In 1987, Kuo and Fuchs proposed the branch-and-bound approach and the heuristic polynomial time approximation approach (Kuo S-Y. and Fuchs W. K. 1987). In 1987, Wey and Lombardi proposed another approach which determines both repairability/unrepairability and a repair solution (Wey C-L. and Lombardi F. 1987). In 1990, Huang et al. proposed the faulty line covering approach and the effective coefficient approach (Huang W. K., Shen Y-N., and Lombardi F. 1990). The problem is proved to be NP-complete (Kuo S-Y. and Fuchs W. K. 1987). No parallel algorithm has been reported.

In this Chapter we propose a parallel algorithm based on the hysteresis McCulloch-Pitts neuron model for solving the spare allocation problem. First we verified our parallel algorithm through solving Huang's problems in (Huang W. K., Shen Y-N., and Lombardi F. 1990). Then we created larger problems where the faulty cells are generated randomly and the size of the fault maps is varied from 20×20 to 200×200. The simulation results are also shown in this Chapter.

8.2 NEURAL REPRESENTATION AND MOTION EQUATIONS

Fig. 8-2a shows the system representation for the spare allocation problem of the 8×8 fault map in Fig. 8-1. Eight processing elements are used to describe on which rows (columns) the spares per row (column) should be allocated. Total 16 processing elements are used in this problem where $V1_i$ represents the output of the i-th row processing element and $V2_i$ represents the output of the i-th column processing element. Generally 2n processing elements are used to represent n rows and n columns for the $n \times n$ fault map problems. The number of processing elements with the nonzero output among the n row processing elements is Ra where Ra is the number of the given spares per row. The number of processing elements with the nonzero output among the n column processing elements is Ca where Ca is the number of the given spares per column.

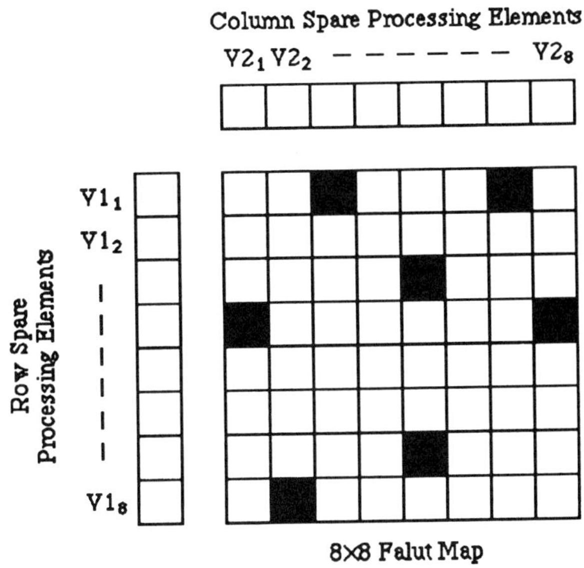

Fig. 8-2 (a) Neural representation of Fig. 8-1

For example, as shown in Fig. 8-2b, the output of two processing elements among eight row processing elements should be nonzero and the output of two processing elements among eight column processing elements should be nonzero, where the black squares indicate the nonzero output of the processing elements and the white squares indicate the zero output of the processing elements. The nonzero output of a processing element means that a spare should be allocated on the corresponding row or column. Fig. 8-2b shows that the spares per row are allocated on the first row and the fourth row, and the spares per column are allocated on the second column and the fifth column. This spare allocation repairs all the faulty cells.

In this problem, each faulty cell in the fault map F must be repaired by a spare per row or a spare per column, or both of them. Therefore it is necessary to allocate a spare per row on the i-th row if a faulty cell of the i-th row is not repaired by a spare per column. This condition is given by:

$$\sum_{k=1}^{n} f_{ik} \cdot (1 - V2_k) \qquad (8.1)$$

This condition is nonzero if a faulty cell of the i-th row is not repaired by a spare per column. In other words, it is nonzero if the output $V2_k$ of the k-th column processing element is zero for the nonzero fault map element f_{ik}.

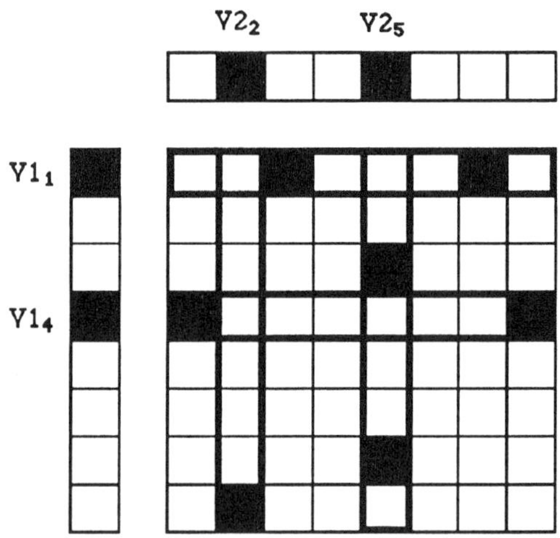

Fig. 8-2 (b) Solution of the problem in Fig. 8-1

Similarly it is also necessary to allocate a spare per column on the i-th column if a faulty cell of the i-th column is not repaired by a spare per row. This condition is given by:

$$\sum_{k=1}^{n} f_{ki} \cdot (1 - V1_k) \qquad (8.2)$$

This condition is nonzero if a faulty cell of the i-th column is not repaired by a spare per row. In other words, it is nonzero if the output $V1_k$ of the k-th row processing element is zero for the nonzero fault map element f_{ki}.

The motion equation of the i-th row processing element for the n×n fault map problem with Ra spares per row is given by:

$$\frac{dU1_i}{dt} = -A\left(\sum_{k=1}^{n} V1_k - Ra\right) + B\left[\sum_{k=1}^{n} f_{ik}\cdot(1-V2_k)\right]\cdot(1-V1_i) \qquad (8.3)$$

where $U1_i$ is the input of the i-th row processing element. The first term (A-term) in Eq. (8.3) forces the output of Ra row processing elements to be nonzero where the spares per row are allocated. The second term (B-term) encourages the output of the i-th row processing element to be nonzero if a faulty cell of the i-th row is not repaired by a spare per column and the output $V1_i$ of the i-th row processing element is zero. A and B are constant coefficients.

Similarly the motion equation of the i-th column processing element for the n×n fault map problem with Ca spares per column is given by:

$$\frac{dU2_i}{dt} = -A\left(\sum_{k=1}^{n} V2_k - Ca\right) + B\left[\sum_{k=1}^{n} f_{ki}\cdot(1-V1_k)\right]\cdot(1-V2_i) \qquad (8.4)$$

where $U2_i$ is the input of the i-th column processing element.

The states of the 2n processing elements for the n×n fault map problem can be updated synchronously or asynchronously using Eq. (8.3) and Eq. (8.4). In the synchronous system, the states of all processing elements are updated simultaneously. In the asynchronous system, the states of all processing elements are asynchronously updated. In this Chapter, the synchronous parallel system is simulated on a sequential machine where the synchronous parallel system can be performed on maximally 2n processors.

The developed simulator uses the following parameter set of the coefficients A and B, UTP, LTP, U_max, and U_min: t=0, A=B=1, UTP=5, LTP=-5, U_max=40, and U_min=-160. First we used the simulator for solving Huang's problems in (Huang W. K., Shen Y-N., and Lombardi F. 1990). Then we created larger problems where the faulty cells are generated randomly and the size of the fault maps is varied from 20×20 to 200×200.

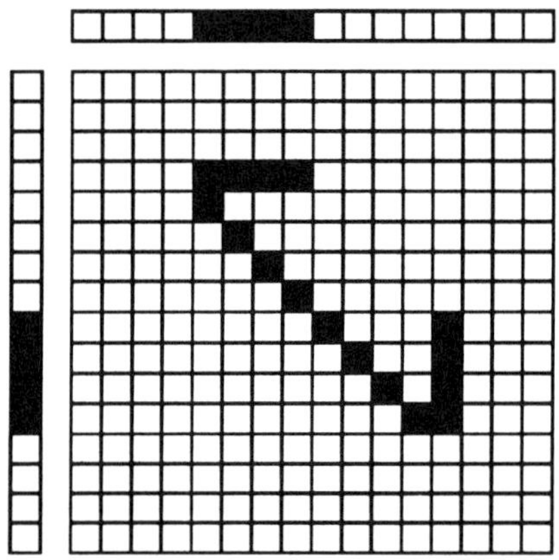

Fig. 8-3 16 x 16 fault map problem with Ra=Ca=4

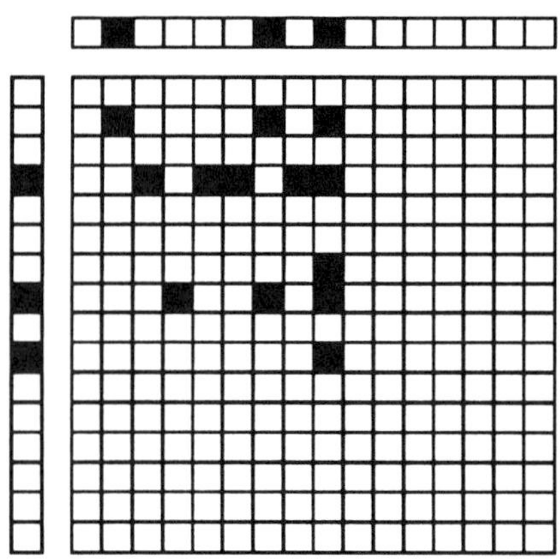

Fig. 8-4 16 x 16 fault map problem with Ra=Ca=3

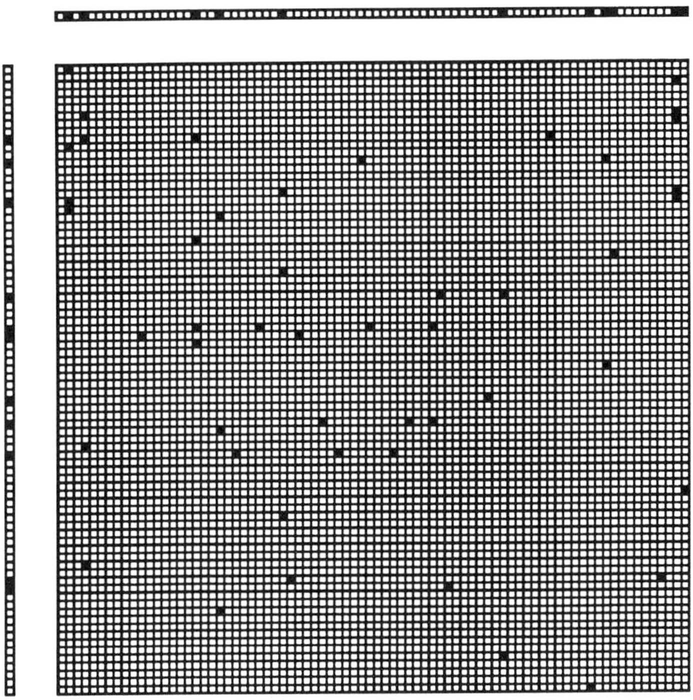

Fig. 8-5 80 x 80 fault map problem with Ra=Ca=11

Fig. 8-3 through 8-5 show the simulation conditions for these problems and the optimum solutions where all the faulty cells are repaired by spares respectively. The algorithm found several optimum solutions from the different initial values of $Ux_i(t)$. Fig. 8-6 shows the simulation results where the number of spares per row and spares per column are varied for the same fault map as that of Fig. 8-4. The simulator discovered that 3 spares per row and 1 spare per column are minimum to repair all the faulty cells in this fault map.

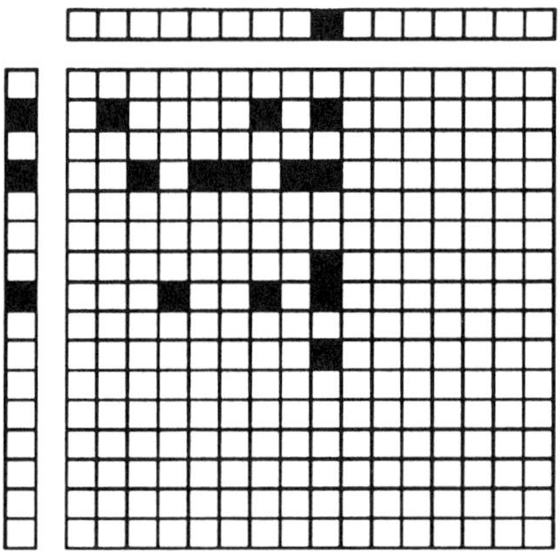

Fig. 8-6 Optimum solution of the problem in Fig. 8-4

Table 8-1 shows the average number of iteration steps and the frequency of the convergence to the optimum solution where 100 simulation runs were performed for each one of 17 problems. See our paper (Funabiki and Takefuji 1991) for the detail of 17 problems and the simulation results. For each one of simulation runs, the different initial values were given. Fig. 8-7 shows the relationship between the frequency and the number of iteration steps to converge to the optimum solution for the 40 x 40 fault map problem. The simulation results show that the average number of the iteration steps and the frequency of the convergence to the optimum solution are not determined by the problem size but by the degree of the difficulty in the problem. We conclude that our algorithm can approximate solutions to the spare allocation problems with at least small size of fault maps in a nearly constant time with O(n) processors.

Table 8-1

Fault Map Problem Size	Average Iteration Steps to Optimum Solution	Convergence Frequency to Optimum Solution
16 × 16 (Fig. 8-3)	430.1	96%
16 × 16 (Fig. 8-4)	79.9	90%
80 × 80 (Fig. 8-5)	279.7	85%

8.3 REFERENCES

Benevit C. A. et al., (1982), A 256K Dynamic Random Access Memory, IEEE J. Solid-State Circuits, Vol. SC-17, No.5, pp.857-862.

Carlson R. O. and Neugebauer C. A., (1986), Future Trends in Wafer Scale Integration, Proc. IEEE, Vol.74, No.12, pp.1741-1752.

Cenker R. P. et al., (1979), A Fault-Tolerant 64K Dynamic Random-Access Memory, IEEE Trans. on Electric Devices, Vol. ED-26, No.6, pp.853-860.

Day J. R., (1985), A Fault-Driven, Comprehensive Redundancy Algorithm, IEEE Design&Test, Vol.2, pp.35-44.

Evans R. C., (1981), Testing Repairable RAMs and Mostly Good Memories, Proc. Int. Test Conf., pp.49-55.

Fitzgerald B. F. and Thoma E. P., (1980), Circuit Implementation of Fusible Redundant Addresses on RAMs for Productivity Enhancement, IBM J. Res. Develop., Vol.24, No.3, pp.291-298.

Funabiki N. and Takefuji Y., (1991), A parallel algorithm for spare allocation problems, to appear in IEEE Trans. on Reliability.

Haddad R. W. and A. T. Dahbura A. T., (1987), Increased Throughput for the Testing and Repair of RAMs with Redundancy, Proc. IEEE ICCAD, pp.230-233.

Hayasaka Y. , Shimotori K., and Okada K., (1982), Testing System for Redundant

Memory, Proc. Int. Test Conf., pp.240-244.

Huang W. K., Shen Y-N., and Lombardi F., (1990), New Approaches for the Repairs of Memories with Redundancy by Row/Column Deletion for Yield Enhancement, IEEE Trans. on Computer-Aided Design, Vol. 9, No. 3, pp.323-328.

Kuo S-Y. and Fuchs W. K., (1987), Efficient Spare Allocation for Reconfigurable Arrays, IEEE Design&Test, Vol.4, pp.24-31.

Lee K. C. and Takefuji Y, (1992), Finding a knight's tour using neural networks, IEEE Trans. on System, Man, and Cybernetics.

McCulloch W. S. and Pitts W. H., (1943), A logical calculus of ideas immanent in nervous activity, Bulletin of Mathematical Biophysics, 5, 115.

Schuster S. E., (1978), Multiple Word/ Bit Line Redundancy for Semiconductor Memories, IEEE J. Solid-State Circuits,Vol. SC-13, No.5, pp.698-703.

Stapper C. H., McLaren A. N. , and Dreckmann M., (1980), Yield Model for Productivity Optimization of VLSI Memory Chips with Redundancy and Partially Good Product, IBM J. Res. Develop., Vol.24, No.3, pp.398-409.

Sud R. and Hardee K. C., (1980), 16-K static RAM takes new route to high speed, Electronics, pp.117-123, Sep. 11, 1980.

Tarr M., Boudreau D., and Murphy R., (1984), Defect analysis system speeds test and repair of redundant memories, Electronics, pp.175-179.

Wey C-L. and Lombardi F., (1987), On the Repair of Redundant RAM's, IEEE Trans. on Computer-Aided Design, Vol. CAD-6, No.2, pp.222-231.

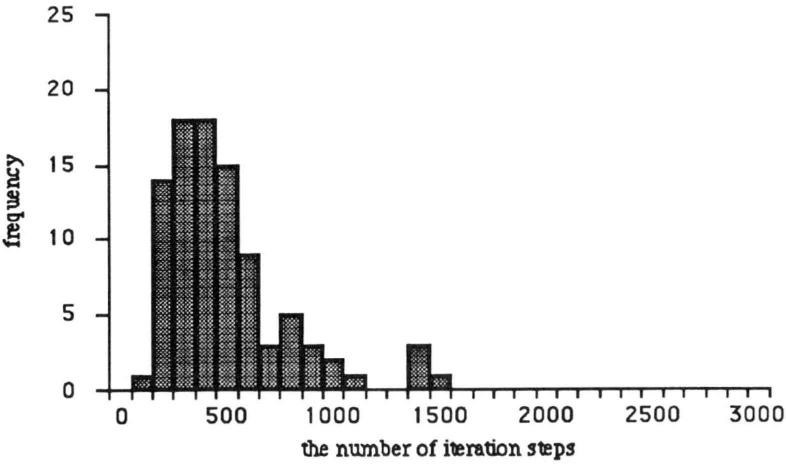

Fig. 8-7 Relationship between the number of iteration steps and the frequency for the 40 x 40 fault map problem

8.4 EXERCISES

1. Calculate the energy function E for the spare allocation problems from Eqs. (8. 3) and (8. 4).
2. Write a subroutine to generate a faulty cell pattern with a specified probability of a faulty cell occurring.
3. Generate a faulty cell pattern and solve it with your simulator.
4. The proposed motion equations in Eqs. (8. 3) and (8. 4) find a spare allocation pattern by using exact Ra spares per row and exact Ca spares per column. However, it is actually sufficient to find a spare allocation pattern by using Ra or less spares per row and Ca or less spares per column. How should the motion equations be modified in order to satisfy this condition ?
5. Modify your simulator with the modified motion equations and run it. Summarize the simulation results.

Chapter 9

SORTING AND SEARCHING

A new neural network parallel algorithm for sorting problems is introduced in this Chapter. The proposed algorithm using $O(n^2)$ processors requires two and only two steps, not depending on the problem size, while the conventional parallel sorting algorithm using O(n) processors proposed by Leighton needs the computation time O(log n). The similar parallel algorithm is introduced for solving string search problems.The proposed algorithm uses $O(m \times n)$ processors where n is the length of a text and m is the length of a pattern. It requires two and only two iteration steps to find a pattern in the text, while the best existing parallel algorithm needs the computation time O(*loglog n*). This Chapter is based on a paper published in IEEE Trans. on Circuits and Systems (Takefuji and Lee 1990a) and a paper published in Journal of Neural Network Computing (Takefuji and Lee 1990).

9.1 INTRODUCTION

Sorting is one of the most fundamental operations in computer science and engineering. The conventional sequential algorithm based on comparisons of pairs of elements must have complexity: O(nlog n). In 1968 Batcher introduced a parallel sorting algorithm with time complexity $O(\log^2 n)$ using $2^{k-2}k(k+1)$ comparators where $n=2^k$ is the number of unsorted elements (Batcher 1968). Leighton provided a parallel algorithm using n processors with O(log n) time (Leighton 1984). Alon and Azar have claimed $\Theta(\log n/\log (1+p/n))$ algorithms using p processors (Alon and Azar Dec. 1988). In this Chapter we first introduce a parallel algorithm based on the McCulloch-Pitts neuron model for sorting. Secondly a string search algorithm is given. From the neural computing viewpoint, sorting and string searching are very similar operations.

A string search problem is to find all occurrences of a given m-character pattern in a given n-character text and it is also one of the fundamental operations in information system and computer science. One of the best known string search algorithms was proposed by Boyer-Moore (Boyer and Moore 1977). Their algorithm performs in $O(n/m)$ on the average case. However, the performance of their algorithm is poor for short patterns when m is small, and the worst case of their algorithm is $O(n+rm)$ where r is the number of matches found (Smit 1982) (Baeza-Yates 1989). The computation time of Knuth's algorithm (Knuth et al. 1977) is given by $O(n+m)$ in the worst case. A few parallel algorithms for solving string search problems have been proposed in the last decade. In 1984 Galil proposed two optimal parallel algorithms where one algorithm requires O(k) time with $p=n^{1+1/k}$ and the other O(logn/loglog n) time with p=n (Galil 1984). In 1985 Vishkin proposed an O(n/p) time parallel algorithm with $p \leq n/\log n$ (Vishkin 1985). Berkman et al. presented an O(loglog n) time parallel algorithm in 1989 (Berkman et al. 1989). String-search VLSI circuits have been investigated by Hirata (Hirata et al. 1988), Foster (Foster and Kung 1980), and Wade (Wade et al. 1987).

9.2 SORTING

In this Section a new parallel distributed algorithm to sort a list of n unsorted elements is presented. The algorithm using $O(n^2)$ processors requires two and only two iteration steps regardless of the size of the problem. N-1 positive integers, N_1, N_2, ..., N_{n-1}, and 0 (zero) which is a dummy number are given where the subscript i indicates the location of the register i which contains the number N_i. The goal of sorting is to find a permutation $(\pi_1, \pi_2, \ldots, \pi_{n-1})$ such that $0 < N_{\pi_1} < N_{\pi_2} < \ldots < N_{\pi_{n-1}}$. In our new sorting an n x n neural network array is provided where each row and column corresponds to the location of a register and the position of permuted order respectively. The n x n neural array actually represents the directed adjacency matrix where one and only one neuron in the *i*th row (i=1,...,n) will be fired in order to determine the sorting order between N_{π_i} and $N_{\pi_{i+1}}$. Note that $N_{\pi_{i+1}}$ must be greater than N_{π_i} but it must be the nearest number to N_{π_i}.

The predefined energy function based on our neural representation is given by:

$$E = \sum_{X=1}^{n} \sum_{Y \neq X}^{n} \frac{A}{(N_Y - N_X)} \sum_{i \neq X}^{n} f(N_Y, N_X) V_{Xi} V_{XY} - \sum_{X=1}^{n} \sum_{Y \neq X}^{n} \frac{A}{(N_Y - N_X)} V_{XY} \tag{9.1}$$

Instead of explaining Eq. (9.1) it is much easier to understand the behavior of the neural network using the motion equation of the *XY*th neuron. The motion equation of the neuron in the *Y*th column and in the *X*th row is given by considering the necessary and sufficient condition of the sorted state.

It is given by:

$$\frac{dU_{XY}}{dt} = - \frac{A}{N_Y - N_X} \left(\sum_{i \neq X}^{n} f(N_Y, N_i) V_{Xi} - 1 \right) \tag{9.2}$$

$$\text{where } f(L,R) = \begin{cases} 0 & \text{if } L \leq R \\ 1 & \text{otherwise} \end{cases}$$

It is very important that all of the initial values at t=0 for U_{ij} (i=1,...,n) (j=1,...,n) are set to small negative numbers. In Eq. (9.2) the term $(\Sigma f(N_Y, N_i) V_{Xi} - 1)$ forces one and only one neuron to fire per row. When N_Y is greater than N_X, Eq. (9.2) at t=0 becomes positive because V_{ij} (i=1,...,n) (j=1,...,n) are zeros. The function of

$f(N_Y,N_i)$ plays a key role in our method to determine which permutation connection should remain and which should be removed. In other words, only if N_Y is greater than N_X and the smallest number among N_i (i=1,...,n-1), the term $\Sigma f(N_Y,N_i)V_{Xi}$ is zero so that Eq. (9.2) will be still positive, otherwise it will be negative. The $A/(N_Y-N_X)$ term and the function $f(N_Y,N_i)$ in Eq. (9.2) will guarantee the state of the system to converge to the global minimum. In order to achieve the global minimum convergence, remember all of the initial values for the inputs U_{ij} (i=1,...,n) (j=1,...,n) must be set to small negative numbers.

Proof:

Consider the neuron in the Yth column and in the Xth row. The initial values of U_{ij} (i=1,...,n) (j=1,...,n) are negative at t=0 so that all of the V_{ij} are zeros, because of the function of binary neurons. If all of the V_{ij} are zeros, then Eq. (9.2) must be positive as long as the value of (N_Y-N_X) is positive. If the value of (N_Y-N_X) is negative, Eq. (9.2) will be negative so that the neuron is not fired. We can conclude that at t=0 the negative initial values of U_{ij} (i=1,...,n) (j=1,...,n) forces the neurons whose number is greater than N_X to fire. After the first iteration, one and only one neuron among fired neurons whose number must be not only greater than N_X but also be the smallest number is forced to remain fired. Because the product of $f(N_Y,N_i)$ and V_{Xi} (i=1,...,n) in Eq. (9.2) enables the neuron in the *Y*th column and in the *X*th row to unfire as long as $N_Y \geq N_i$ is satisfied. If the N_Y is the smallest number, then the term $\Sigma f(N_Y,N_i)V_{Xi}$ will be zero so that the neuron is forced to remain fired. Consequently our method only requires two iteration steps to converge to the global minimum as long as the appropriate number is assigned to Δt where it must be greater than the absolute values of the initial U_{ij} (i=1,...,n) (j=1,...,n) at time t=0. **Q.E.D.**

Our simulator using Eq. (9.2) is based on the first order Euler method;

$$U_{XY}(t+\Delta t)=U_{XY}(t)+\Delta U_{XY}\Delta t \qquad (9.3)$$

where ΔU_{XY} is dU_{XY}/dt. Fig. 9-1a shows the relationship between the size of the problems and the number of iterations for the global minimum convergence where Δt was varied 10^{-4} to 10^{-6} and the initial values of U_{XY} were set to 1/70000. Unsorted positive numbers were generated randomly. When the absolute values of the initial U_{ij} (i=1,...,n) (j=1,...,n) are greater than Δt, the number of iterations requires more than two. When Δt is greater than those, the state of the system always converges to the global minimum with two and only two iteration steps.

In order to remove the expensive term $A/(N_Y-N_X)$ from Eq. (9.2) which must use the variable conductance devices, the following motion equation is created.

$$\frac{dU_{XY}}{dt} = -f(N_Y,N_X)\left(\sum_{j\neq X}^{n} f(N_Y,N_j)V_{Xj}-1\right). \tag{9.4}$$

It is very important to notice that Eq. (9.4) does not require the variable conductance devices. Fig. 9-1b shows the relationship between the number of iterations and the problem size based on Eq. (9.4). The number of iteration steps can be given by the following formula:

$$\text{The number of iteration steps} = \text{Integer}\left[\frac{|U_{ij}(0)|}{\Delta t}\right] + 2 \tag{9.5}$$

where $U_{ij}(0)$ is the initial value at time t=0 (i =1,...,n and j =1,...,n).

The sorting neural network based on Eq. (9.4) needs n^2 McCulloch-Pitts binary neurons and n(n-1)/2 comparators to sort n-1 unsorted elements. Fig. 9-2 shows a simple example to sort 9 numbers (44, 3, 62, 33, 57, 91, 37, 82, 13) using a 10 x 10 neural network array. The simulation result indicates $0<3<13<33<37<44<57<62<82<91$. Black and white squares indicate positive and negative respectively where the linear dimension of each square is proportional to the value of ΔU_{Xi}, U_{Xi}, V_{Xi}.

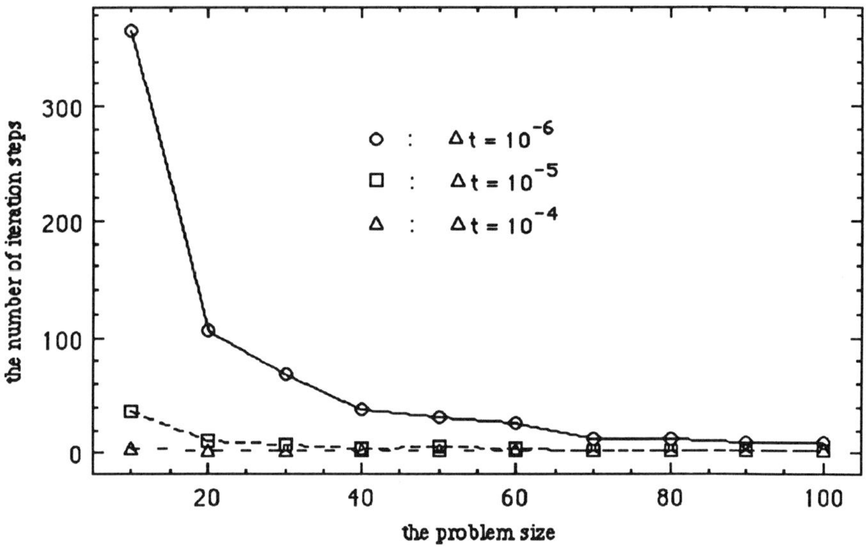

Fig. 9-1 (a) Relationship between the number of iteration steps and the problem size

We have investigated the problem size up to n=1000. All of our results showed that the computation time is constant, namely two-step when the following condition is satisfied:

$U_{ij}(0) < \Delta t$ (i =1,...,n and j =1,...,n).

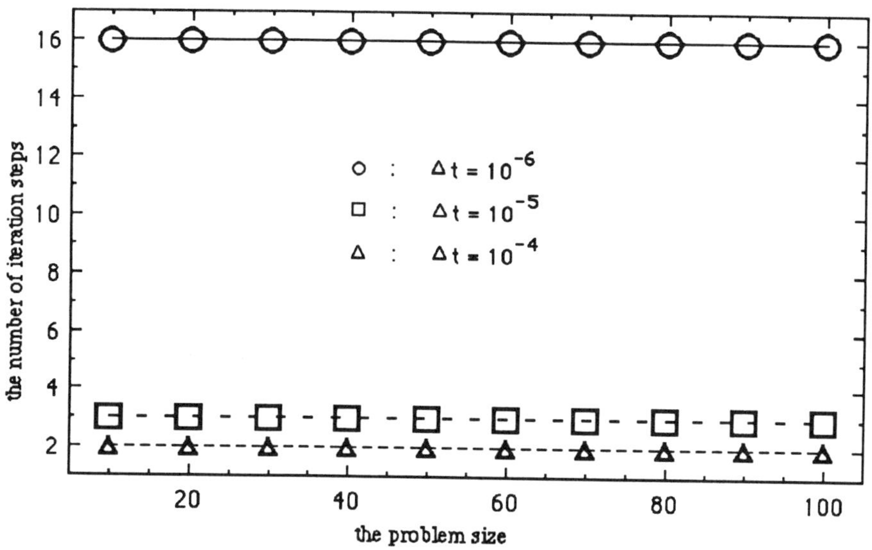

Fig. 9-1 (b) Relationship between the number of iteration steps and the problem size

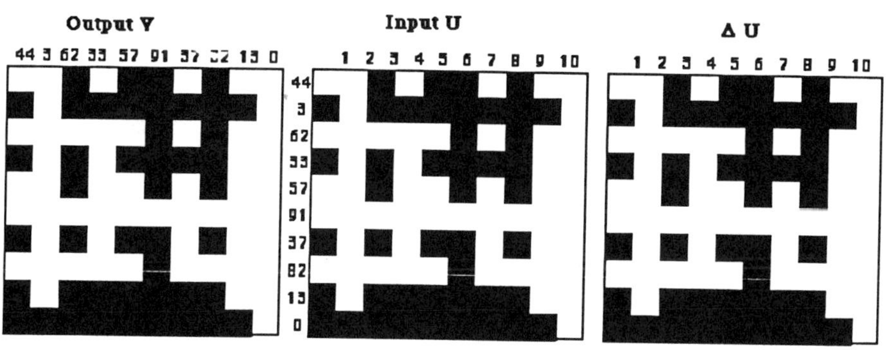

Fig. 9-2 (a) The state of the system after the first iteration step

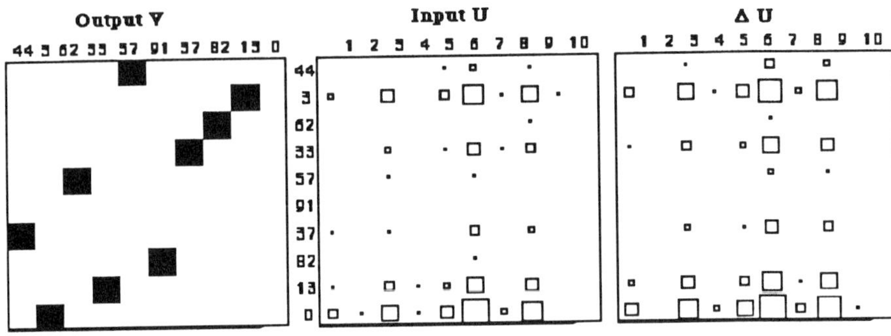

Fig. 9-2 (b) The final state of the system after the second iteration step

9.3 SEARCHING

The string search problem is to find all occurrences of a given m-character pattern in a given n-character text and it is one of the fundamental operations in information system and computer science. An m x (n-m+1) neural network array is provided for solving a string search problem where row and column correspond to the location of the pattern register and that of the text register respectively. U_{iX} and V_{ix} represent an input and an output of the iXth neuron respectively where the subscript i indicates the ith register which contains the ASCII value of the ith character in the pattern, N_{pi}. In the same manner, the second subscript X indicates the Xth register which contains the ASCII value of the Xth character in the text, N_{tX}. If more than one pattern exist in the given text, the corresponding iXth neuron will be fired, that is, V_{ix} =1. For example, Fig. 9-3 shows the neural network representation for searching the 4-character pattern "name" in the 18-character text "My name is Neuron.". Fired neurons locate the string of the searching pattern.

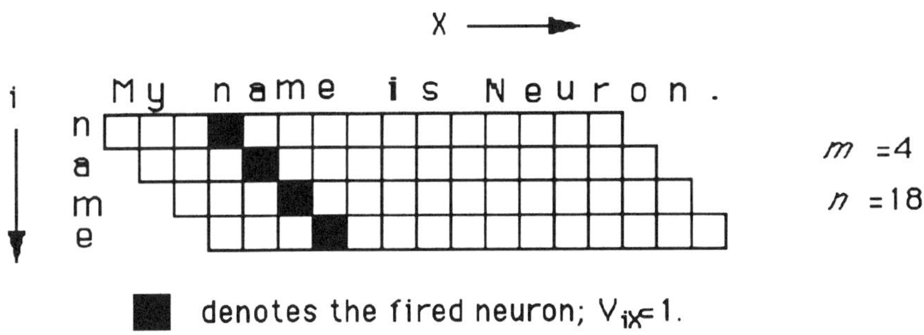

Fig. 9-3 Neural network representation

The motion equation of the *iX*th neuron is given by:

$$\frac{dU_{ix}}{dt} = -A \cdot g(N_{pi}-N_{tX}) - B \cdot g\left(\sum_{j=1}^{m} V_{j\,X-i+j} - m\right) \tag{9.6}$$

where the function $g(x)$ is -1 if x=0, 1 otherwise. A and B are coefficients parameters and m is the length of a pattern. In Eq. (9.6), the first term performs the excitatory forces as long as the *i*th character in the pattern is the same as the *X*th character in the text. If $N_{pi}=N_{tX}$ then V_{iX} will be 1, else the first term will act as the inhibitory force. The second term will be the excitatory force when diagonally consecutive m neurons including the *iX*th neuron are all fired. If not, the second term will discourage the *iX*th neuron to fire.

The first order Euler method was used for the numerical simulation of Eq. (9.6) and each simulation run was terminated when the patterns were found in the given text. Initial values of neurons U_{iX} for i=1,...,m and X=1,...,n were assigned to the same negative value. Fig. 9-4a shows the state of the neurons after the first iteration step. At this stage, every *iX*th neuron whose N_{pi} equals to N_{tX} is fired because the first A term of Eq. (9.6) is positive and the second term is zero. After the second step, only diagonally consecutive m neurons among fired neurons are forced to remain fired as shown in Fig. 9-4b. At the second step, the second term of Eq. (9.6) encourages the exact matching patterns to remain while the first term is zero.

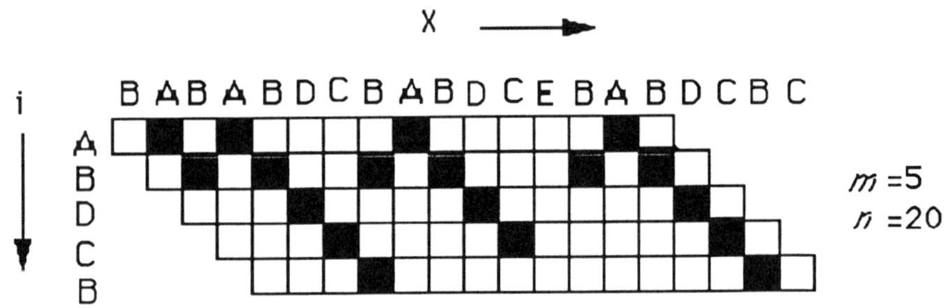

Fig. 9-4 (a) The state of the system after the first iteration step

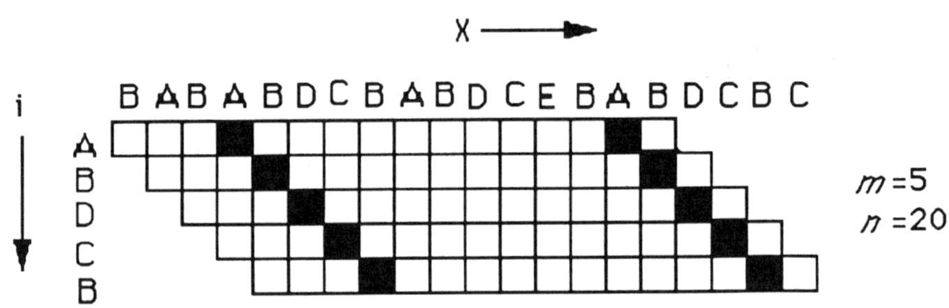

Fig. 9-4 (b) The final state of the system after the second iteration step

A large number of simulation runs were performed with various length of texts and patterns on Macintosh SE/30. Fig. 9-5 shows the simulation result of the text length 100. The characters of the text and the pattern in Fig. 9-5 were randomly generated. In Fig. 9-5, the pattern "cdac" is located at two places in the 100-length text.

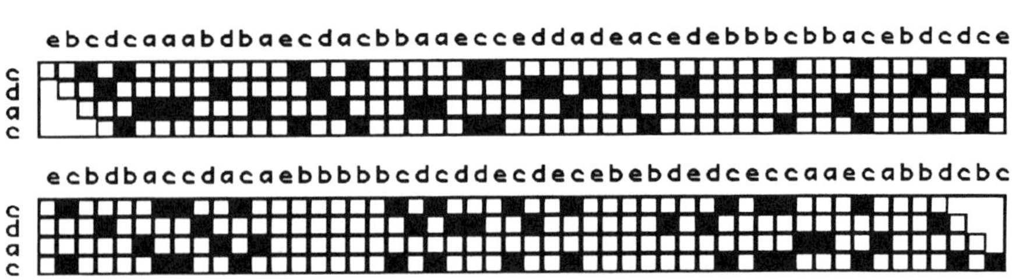

Fig. 9-5 (a) The state of the system after the first iteration step

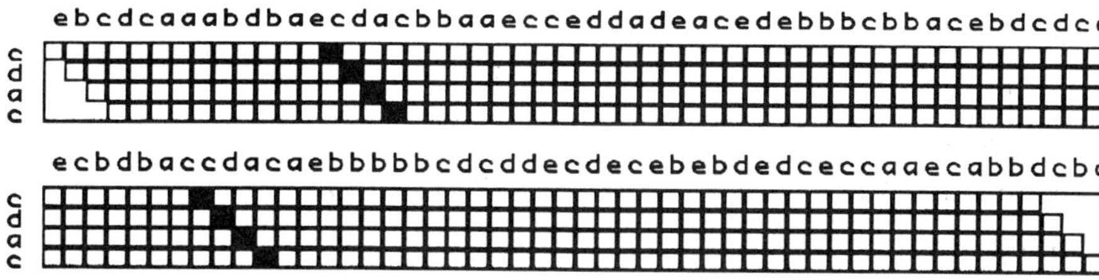

Fig. 9-5 (b) The final state of the system after the second iteration step

9.4 REFERENCES

Alon N., and Azar Y., (Dec. 1988), The average complexity of deterministic and randomized parallel comparison-sorting algorithms, SIAM J. Computing, Vol. 17, No. 6.

Batcher K. E., (1968), Sorting networks and their applications, Proc. of the Spring Joint Computer Conf., Vol. 32, AFIPS Press.

Leighton F. T., (1984), Tight bounds on the complexity of parallel sorting, Proc. of the ACM sym. on Theory of Computing, pp.71-80, ACM.

Boyer R. S.,and Moore J. S., (1977), A fast string searching algorithm, Communications of the ACM, **20**, 762-772.

Smit G. D., (1982), A comparison of three string matching algorithms, Software Practice and Experience, **12**, 57-66.

Baeza-Yates R. A., (1989), Improved string searching, Software Practice and Experience, **19**, 257-271.

Knuth D. E., Morris J. H. Jr., and Pratt V. B., (1977), Fast pattern matching in strings, SIAM J. Computing, **6**, 323-350.

Galil Z., (1984), Optimal parallel algorithms for string matching, Proc. of 16th ACM symposium on Theory of Computing, 240-248.

Vishkin U., (1985), Optimal parallel pattern matching in strings, Proc. of 12th ICALP, Lecture Notes in Computer Science 194, Springer-Verlag, 497-507.

Berkman O., Breslauer D., Galil, Schieber B., and Vishkin, (1989), Highly parallelizable problems, Proc. of 21st ACM Symposium on Theory of Computing.

Hirata M., Yamada H., Nagai H., and Takahashi K., (1988), A versatile data string-search VLSI, IEEE J. of Solid-State Circuits, 23, 2, 329-335.

Foster M. J., and Kung H. T., (1980), The design of special-purpose VLSI chips, IEEE Computer, 13, 1, 26-40.

Takefuji Y. and Lee K. C, (1990), A super parallel sorting algorithm based on neural networks, IEEE Trans. on Circuits and Systems, 37, 11, 1425-1429.

Takefuji Y. and Lee K. C., (1990), A two-step parallel sorting algorithm based on neural networks, Journal of Neural Network Computing, 2, 1, 30-32.

Takefuji Y., Tanaka T., and Lee K. C., (1992), to appear in IEEE Trans. on System, Man, and Cybernetics.

Wade J. P., Osler P. J., Zippel R. E., and Sodini C. G., (1987), The MIT database accelerator: 2-K TRIT circuit design, Proc. of VLSI symposium Dig. Tech., 39-40.

Chapter 10

TILING PROBLEMS

A parallel algorithm for tiling with polyominoes is presented in this Chapter. The tiling problem is to pack polyominoes in a finite checkerboard. The algorithm using p x m x n processing elements requires O(1) time where p is the number of different kinds of polyominoes on an m x n checkerboard. The algorithm can be used for placement of components or cells in a very large integrated circuit (VLSI) chip, designing and compacting printing circuit boards, and solving a variety of 2-D or 3-D packing problems. This Chapter is based on a paper published in IEEE Trans. on Neural Networks (Takefuji and Lee 1990).

10.1 INTRODUCTION

The problem of tiling with polyominoes was introduced by S. W. Golomb in 1953 (Gardner 1975). The tiling problem is to pack a checkerboard with polyominoes. Although Klarner (Klarner 1969), Golomb (Golomb 1970), Gardner (Gardner 1969), Klamkin and Liu (Klamkin and Liu 1980), Golomb (Golomb 1989) and others have been working on the problem, few sequential algorithms have been reported. No parallel algorithm has been given during the last three decades. Akiyama et al. (Akiyama et al. 1989) have proposed the first parallel algorithm based on the stochastic neural network using noise to escape from the local minimum. However the quality of the solution drastically degrades with the problem size although the experimented problem size is small. The complexity of solving their 5 x 5 checkerboard problem is to find a single solution among 3.3×10^6 possible candidates.

This Chapter introduces a new deterministic parallel tiling algorithm based on the modified McCulloch-Pitts neuron model. The algorithm packs a checkerboard with polyominoes within O(1) time. The quality of the solution does not degrade with the problem size. The algorithm was verified by more than one thousand simulation runs solving a 7 x 7 checkerboard tiling problem with 10 polyominoes. Without rotation or reflection, the complexity of our problem is to find a single solution among 1.3×10^{14} ($\approx 25^6 \times 36 \times 30 \times 24 \times 21$) possible candidates where there exists one and only one solution in the problem.

10.2 NEURAL REPRESENTATION AND MOTION EQUATIONS

The algorithm uses a three dimensional 10 x 7 x 7 neural network array where the output of the *ijk*th neuron follows:

$$V_{ijk}=1 \text{ if } U_{ijk}>0 \text{ and } U_{ijk}=\max(U_{ixy}) \text{ for } x=1,\ldots,7 \text{ and } y=1,\ldots,7 \qquad (10.1)$$
$$\quad 0 \text{ otherwise}$$

Fig. 10-1 (a)-(j) show ten polyominoes and their markers respectively. A marker is used to locate a polyomino on the checkerboard with respect to its coordinate. In Fig. 10-1 a shaded square not only indicates the coordinate of a marker on the sideboard and but also the coordinate of a fired neuron in a 7 x 7 neural array. A black square on a 7 x 7 neural array (left in Fig. 10-1) means that a neuron on the coordinate is fired and the state of the output is nonzero. White squares means that neurons are not fired on their coordinates. A polyomino requires a two dimensional 7 x 7 neural network array so that this problem can be solved by a three dimensional 10 x 7 x 7 neural network array as shown in Fig. 10-3. One and only one marker neuron is forced to fire per 7 x 7 neural network array. Besides a 7 x 7 neural network array each polyomino needs a tiling sideboard to represent its shape and violation conditions. In other words, the state of the sideboard indicates occupied tiles of a polyomino on a checkerboard where the sideboard is composed of a 7 x 7 binary array where V'_{ijk} indicates the state of the (j,k) coordinate on the *i*th sideboard.

*1

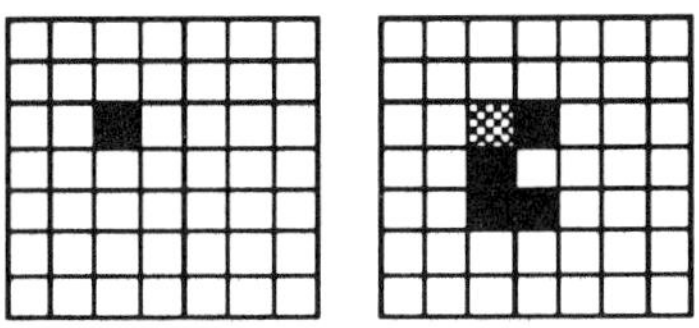

Fig. 10-1 (a) Fired neuron (left) and sideboard with respect to a marker (right)

*2

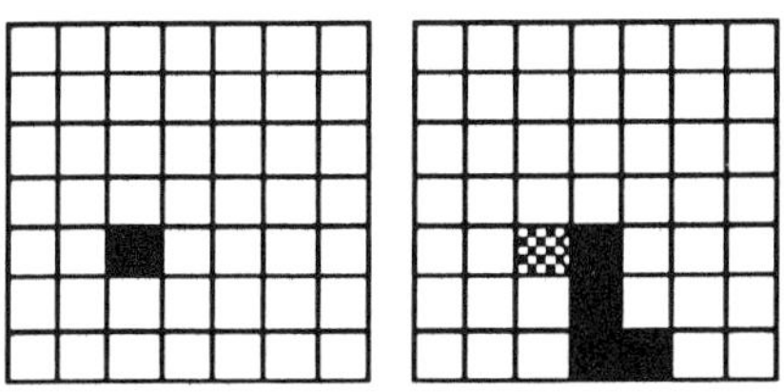

Fig. 10-1 (b) Neuron (left) and sideboard with respect a marker (right)

*3

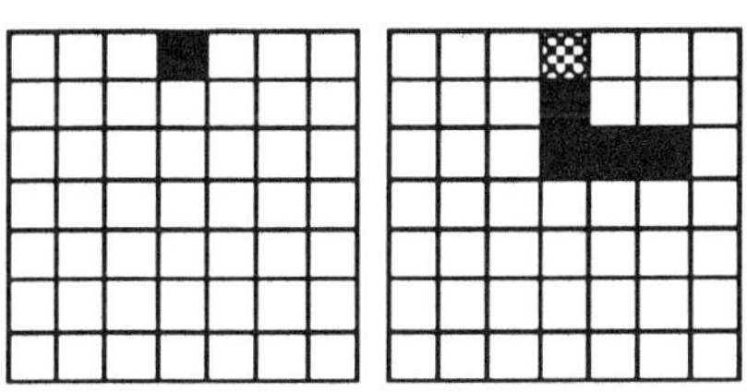

Fig. 10-1 (c) Neuron (left) and sideboard with respect a marker (right)

#4

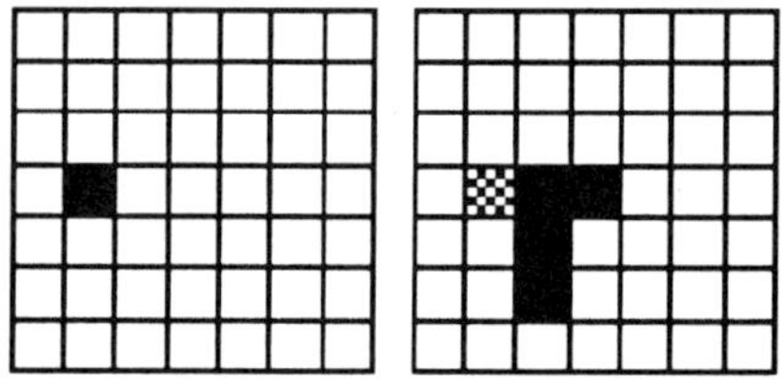

Fig. 10-1 (d) Neuron (left) and sideboard with respect a marker (right)

#5

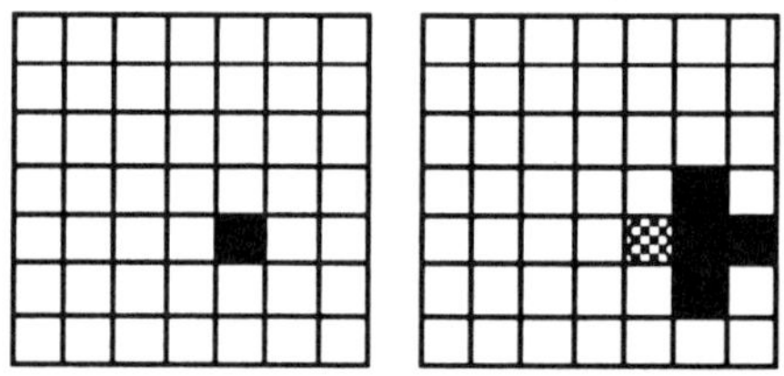

Fig. 10-1 (e) Neuron (left) and sideboard with respect a marker (right)

#6

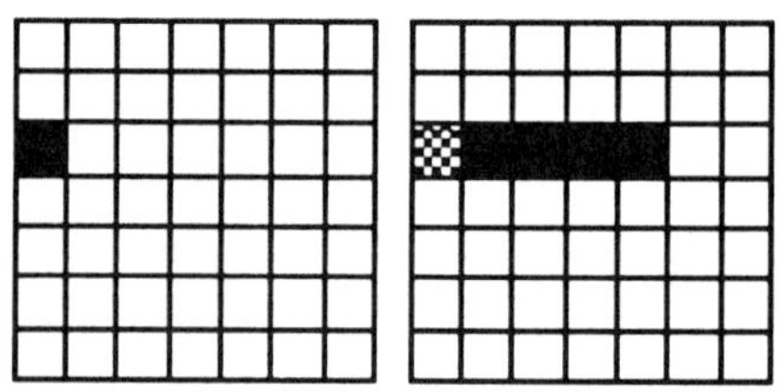

Fig. 10-1 (f) Neuron (left) and sideboard with respect a marker (right)

#7

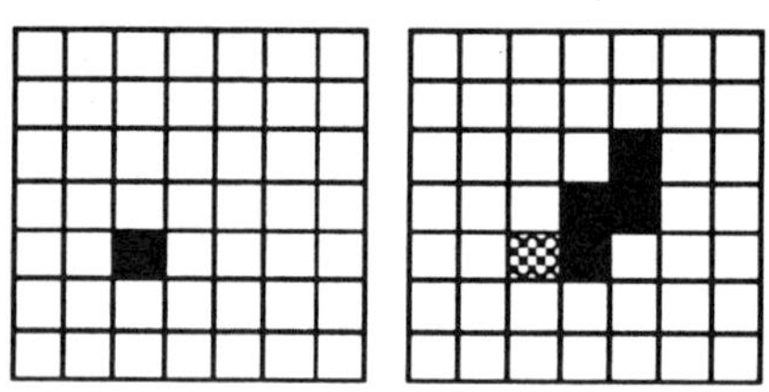

Fig. 10-1 (g) Neuron (left) and sideboard with respect a marker (right)

#8

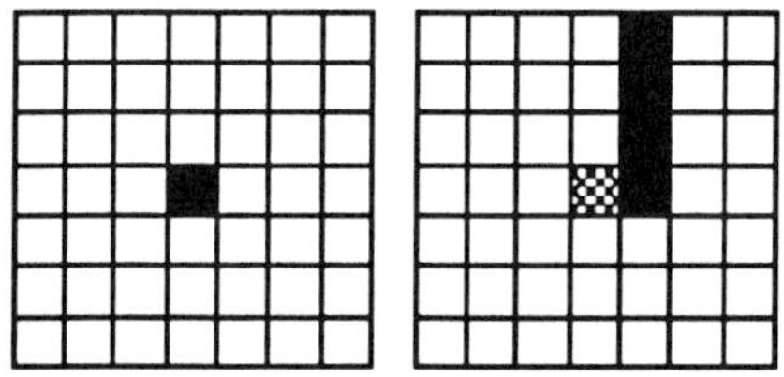

Fig. 10-1 (h) Neuron (left) and sideboard with respect a marker (right)

#9

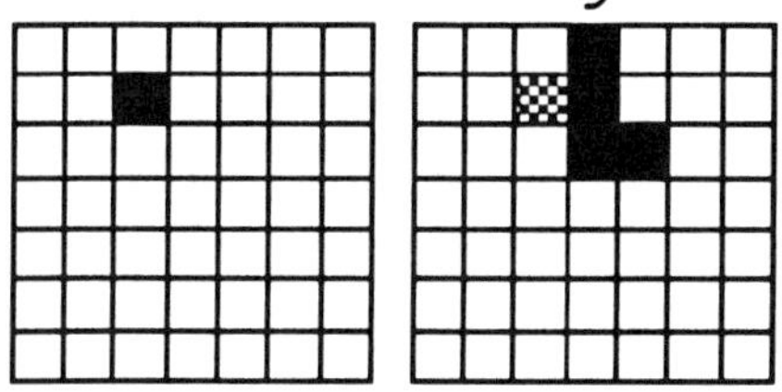

Fig. 10-1 (i) Neuron (left) and sideboard with respect a marker (right)

#10

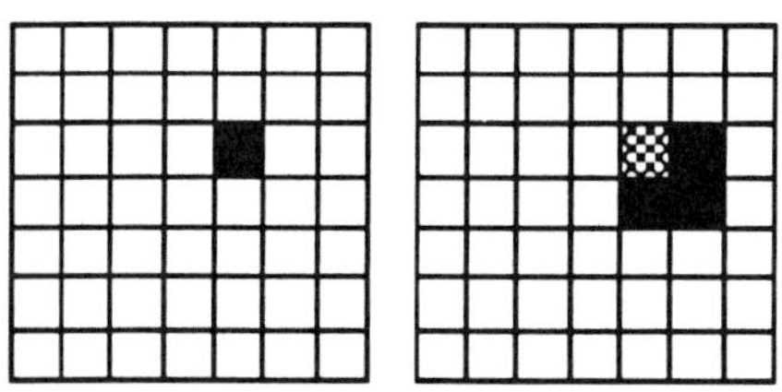

Fig. 10-1 (j) Neuron (left) and sideboard with respect a marker (right)

The motion equation of the *ijk*th neuron is given by:

$$\frac{dU_{ijk}}{dt} = -A\left(\sum_{q=1}^{m}\sum_{r=1}^{n} V_{iqr}-1\right) -B\left(\sum_{q=1}^{l} V_{qjk}-1\right) -Cf(i) +Dh\left(\sum_{q=1}^{m}\sum_{r=1}^{n} V_{iqr}\right) \tag{10.2}$$

where U_{ijk} is the input of the *jk*th neuron on the *i*th neural network array. In Eq. (10.2) the first term forces one and only one neuron to fire in the *i*th polyomino neural network array. The second term forces no two markers to be placed in the same position on the checkerboard. The third term is always inhibitory which describes the overlap violation between polyominoes where the ten violation functions for ten different polyominoes: f(1), f(2), f(3), f(4), f(5), f(6), f(7), f(8), f(9), and f(10) are given by respectively:

$$f(1)=\sum_{\substack{q=1\\q\neq 1}}^{11}\left(V'_{qjk}+V'_{q,j+1,k}+V'_{q,j,k+1}+V'_{q,j,k+2}+V'_{q,j+1,k+2}\right)$$

$$f(2)=\sum_{\substack{q=1\\q\neq 2}}^{11}\left(V'_{qjk}+V'_{q,j+1,k}+V'_{q,j+1,k+1}+V'_{q,j+1,k+2}+V'_{q,j+2,k+2}\right)$$

$$f(3)=\sum_{\substack{q=1\\q\neq 3}}^{11}\left(V'_{qjk}+V'_{q,j,k+1}+V'_{q,j,k+2}+V'_{q,j+1,k+2}+V'_{q,j+2,k+2}\right)$$

$$f(4)=\sum_{\substack{q=1\\q\neq 4}}^{11}\left(V'_{qjk}+V'_{q,j+1,k}+V'_{q,j+2,k}+V'_{q,j+1,k+1}+V'_{q,j+1,k+2}\right)$$

$$f(5)=\sum_{\substack{q=1\\q\neq 5}}^{11}\left(V'_{qjk}+V'_{q,j+1,k}+V'_{q,j+1,k-1}+V'_{q,j+1,k+1}+V'_{q,j+2,k}\right)$$

$$f(6)=\sum_{\substack{q=1\\q\neq 6}}^{11}\left(V'_{qjk}+V'_{q,j+1,k}+V'_{q,j+2,k}+V'_{q,j+3,k}+V'_{q,j+4,k}\right)$$

$$f(7)=\sum_{\substack{q=1\\q\neq 7}}^{11}\left(V'_{qjk}+V'_{q,j+1,k}+V'_{q,j+1,k-1}+V'_{q,j+2,k-1}+V'_{q,j+2,k-2}\right)$$

$$f(8)=\sum_{\substack{q=1\\q\neq 8}}^{11}\left(V'_{qjk}+V'_{q,j+1,k}+V'_{q,j+1,k-1}+V'_{q,j+1,k-2}+V'_{q,j+1,k-3}\right)$$

$$f(9)=\sum_{\substack{q=1\\q\neq 9}}^{11}\left(V'_{qjk}+V'_{q,j+1,k}+V'_{q,j+1,k-1}+V'_{q,j+1,k+1}+V'_{q,j+2,k+1}\right)$$

$$f(10)=\sum_{\substack{q=1\\q\neq 10}}^{11}\left(V'_{qjk}+V'_{q,j+1,k}+V'_{q,j,k+1}+V'_{q,j+1,k+1}\right)$$

Remember that the violation functions are determined by their shapes. For example, f(1) is given by $f(1)=\sum_{\substack{q=1\\q\neq 1}}^{11}\left(V'_{qjk}+V'_{q,j+1,k}+V'_{q,j,k+1}+V'_{q,j,k+2}+V'_{q,j+1,k+2}\right)$ because #1 polyomino is composed of five squares as shown in Fig. 10-2 where the marker has the coordinate (j, k).

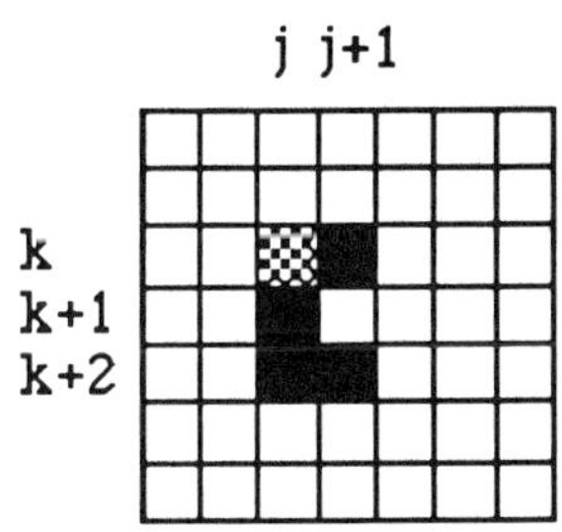

Fig. 10-2 Occupied squares with respect to a marker

Another extra sideboard #11 is used to destroy the isolated space on a checkerboard. Set $V'_{11jk}=1$ if the isolated space is found. The condition of the isolated space is given by:

$$\sum_{q=1}^{10} V'_{qjk}=0,\ \sum_{q=1}^{10} V'_{qj-1,k}=1,\ \sum_{q=1}^{10} V'_{qj+1,k}=1,$$

$$\sum_{q=1}^{10} V'_{qjk-1}=1, \text{ and } \sum_{q=1}^{10} V'_{qjk+1}=1.$$

The third term $-Cf(i)$ in Eq. (10.2) and $-Cf(i)V_{ijk}$ are alternatively used to converge to the global minimum. Related values of V'_{iqr} are set to one if the marker neuron V_{ijk} on the *i*th neural network array is fired where q and r are depending on the polyomino's shape.

For example, set

$V'_{1jk}=V'_{1,j+1,k}=V'_{1,j,k+1}=V'_{1,j,k+2}=V'_{1,j+1,k+2}=1$ if $V_{1jk}=1$ for the #1 polyomino in Fig. 10-1. The last term provides hill-climbing which allows the state of the system to escape from the local minimum and to converge to the global minimum with varying w as shown in step 6 in the proposed procedure. The function h(x) is 1 if x=0, 0 otherwise.

More than one thousand simulation runs were performed to observe the frequency of the system to converge to the global minimum. Table 10-1 shows the result. When w=7 was used, the state of the system always converged to the global minimum within 5000 iteration steps. The relationship between the frequency and the number of iteration steps to converge to the global minimum was observed. The average number of iteration steps for the proposed system to solve the 7 x 7 checkerboard tiling problem without rotation and reflection is 930. Fig. 10-4 shows one and only one solution of the problem without rotation and reflection.

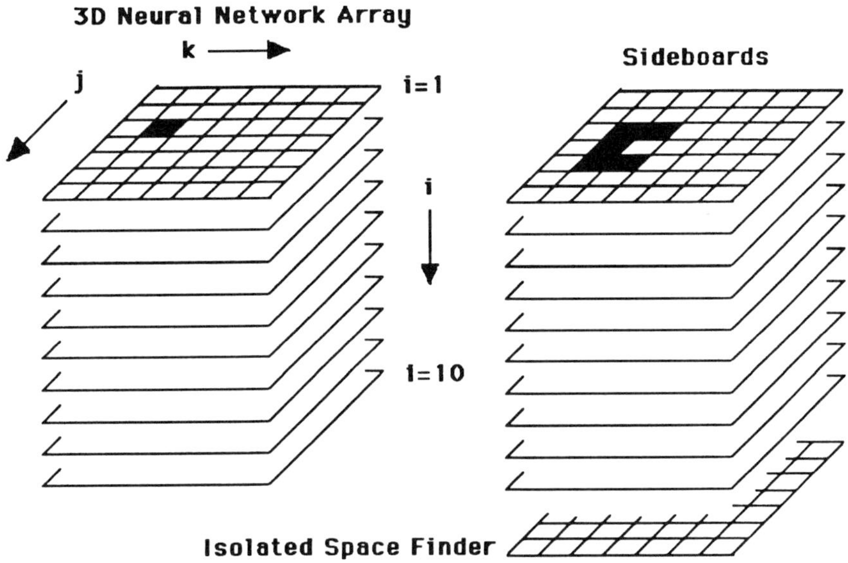

Fig. 10-3 Neural representation and sideboards for 7 x 7 tiling problem

The following procedure describes the proposed algorithm based on the first order Euler method.

0. Set t=0.

1. The initial values of $U_{ijk}(t)$ for i=1,...,10 j=1,...,7 k=1,...,7 are randomized.

2. Evaluate values of $V_{ijk}(t)$ based on the conditional binary function.

$V_{ijk}(t)=1$ if $U_{ijk}(t)>0$ and $U_{ijk}(t)=\max\{U_{ixy}(t)\}$ for x=1,...,7 and y=1,...,7

0 otherwise

3. Values of $V'_{ijk}(t)$ are set to zero.

4. Set appropriate values of $V'_{ijk}(t)$ to one if the mark neuron is fired.

$V'_{1jk}=V'_{1,j+1,k}=V'_{1,j,k+1}=V'_{1,j,k+2}=V'_{1,j+1,k+2}=1$ if $V_{1jk}=1$

$V'_{2jk}=V'_{2,j+1,k}=V'_{2,j+1,k+1}=V'_{2,j+1,k+2}=V'_{2,j+2,k+2}=1$ if $V_{2jk}=1$

$V'_{3jk}=V'_{3,j,k+1}=V'_{3,j,k+2}=V'_{3,j+1,k+2}=V'_{3,j+2,k+2}=1$ if $V_{3jk}=1$

$V'_{4jk}=V'_{4,j+1,k}=V'_{4,j+2,k}=V'_{4,j+1,k+1}=V'_{4,j+1,k+2}=1$ if $V_{4jk}=1$

$V'_{5jk}=V'_{5,j+1,k}=V'_{5,j+1,k-1}=V'_{5,j+1,k+1}=V'_{5,j+2,k}=1$ if $V_{5jk}=1$
$V'_{6jk}=V'_{6,j+1,k}=V'_{6,j+2,k}=V'_{6,j+3,k}=V'_{6,j+4,k}=1$ if $V_{6jk}=1$
$V'_{7jk}=V'_{7,j+1,k}=V'_{7,j+1,k-1}=V'_{7,j+2,k-1}=V'_{7,j+2,k-2}=1$ if $V_{7jk}=1$
$V'_{8jk}=V'_{8,j+1,k}=V'_{8,j+1,k-1}=V'_{8,j+1,k-2}=V'_{8,j+1,k-3}=1$ if $V_{8jk}=1$
$V'_{9jk}=V'_{9,j+1,k}=V'_{9,j+1,k-1}=V'_{9,j+1,k+1}=V'_{9,j+2,k+1}=1$ if $V_{9jk}=1$
$V'_{10jk}=V'_{10,j+1,k}=V'_{10,j,k+1}=V'_{10,j+1,k+1}=1$ if $V_{10jk}=1$

5. Set $V'_{11jk}=1$ if

$$\sum_{q=1}^{10} V'_{qjk}=0,\ \sum_{q=1}^{10} V'_{qj-1,k}=1,\ \sum_{q=1}^{10} V'_{qj+1,k}=1,\ \sum_{q=1}^{10} V'_{qjk-1}=1, \text{ and } \sum_{q=1}^{10} V'_{qjk+1}=1$$

6. Use the motion equation in Eq. (2) to compute $\Delta U_{ijk}(t)$.

if (t mod 10)< w then

$$\Delta U_{ijk}(t) = -A\left(\sum_{q=1}^{m}\sum_{r=1}^{n} V_{iqr}(t)-1\right)-B\left(\sum_{q=1}^{1} V_{qjk}(t)-1\right)-Cf(i)V_{ijk}(t)+\Delta h\left(\sum_{q=1}^{m}\sum_{r=1}^{n} V_{iqr}(t)\right)$$

else

$$\Delta U_{ijk}(t) = -A\left(\sum_{q=1}^{m}\sum_{r=1}^{n} V_{iqr}(t)-1\right)-B\left(\sum_{q=1}^{1} V_{qjk}(t)-1\right)-Cf(i)+\Delta h\left(\sum_{q=1}^{m}\sum_{r=1}^{n} V_{iqr}(t)\right)$$

7. Compute $U_{ijk}(t+1)$ based on the first order Euler method:

$U_{ijk}(t+1)=U_{ijk}(t)+\Delta U_{ijk}(t)$ for i=1,...,10 j=1,...,7 k=1,...,7

8. If $U_{ijk}(t+1)>U_{max}$ then $U_{ijk}(t+1)=U_{max}$
   If $U_{ijk}(t+1)<U_{min}$ then $U_{ijk}(t+1)=U_{min}$

9. Increment t by 1. If t=T terminate this procedure else go to step 2.

Table 10-1 Frequency of convergence to the solution

ω	No. of placed polyominoes	
	<10	10
5	55%	45%
6	21%	79%
7	0%	100%
8	No convergence	

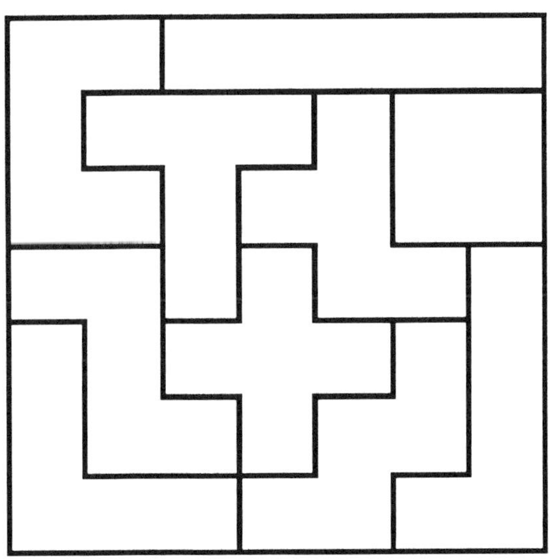

Fig. 10-4 One and only one solution for a 7 x 7 checkerboard problem

10.3 REFERENCES

Akiyama Y., Yamashita A., Kajiura M., and Aiso H., Combinatorial optimization with Gaussian machines, Proc. of the Int. Joint Conf. on Neural Networks, (1989).

Gardner M., (July 1967), Mathematical games, Sci. Amer., 217, 115-116.

Gardner M., (August 1975), Mathematical games, Sci. Amer..

Golomb W. S., (1970), Tiling with sets of polyominoes, J. of Combinarorial Theory, 9, 60-71.

Golomb W. S., (1989), Polyominoes which tile rectangles, J. of Combinatorial Theory, A 51, 117-124.

Klamkin M. S. and A. Liu, (1980), Polyominoes on the infinite checker-board, J. of Combinatorial Theory, A 28, 7-16.

Klarner D. A., (1969), Packing a rectangle with congruent N-ominoes, J. of Combinatorial Theory, 7, 107-115.

Takefuji Y. and Lee K. C., (1990), A parallel algorithm for tiling problems, IEEE Trans. on Neural Networks, 1, 1, 143-145.

10.4 EXERCISES

1. Build a simulator and observe the relationship between the frequency and the number of iteration steps to converge to the global minimum.
2. Observe the behavior of the simulator with varying the value of ω to construct Table 10-1.
3. Develop a motion equation for the following checkerboard problem and build a simulator.
4. Solve the following checkerboard problem.

Chapter 11

SILICON NEURAL NETWORKS

In this Chapter the human brain is first discussed which is largely based on the MS thesis published by Y. Wang in 1991 and analog/digital circuitries for implementing the proposed artificial neural networks are based on two papers published in Electronics Letters in 1990 and published in IEEE Circuits and Devices in 1990.

11.1 INTRODUCTION

The inquisitive spirit is instilled within all human. For much of our past we were groping in the dark, and we still are, for concepts that would help to describe the brain's function. This Section summarizes some known facts and indecisive theories.

Brain Anatomy

The largest and the most important part of the brain is cerebrum. It is cleanly divided down the middle into left and right cerebral hemispheres, and considerably less cleanly front and back into the frontal lobe and the other lobes: the parietal, temporal and occipital in Fig. 11-1. The lobes are not functional regions but convenient anatomical divisions. The outer surface of the cerebrum is called neocortex (or sometimes just cortex). It consists of four to six layers of cells (gray matter) beneath which their axons form pathways (white matter). Neocortex comprises 80 percent of the human brain. It has an area of up to 2500 cm^2 but a thickness of only 1.5 to 3.0 mm. Its surface is wrinkled.

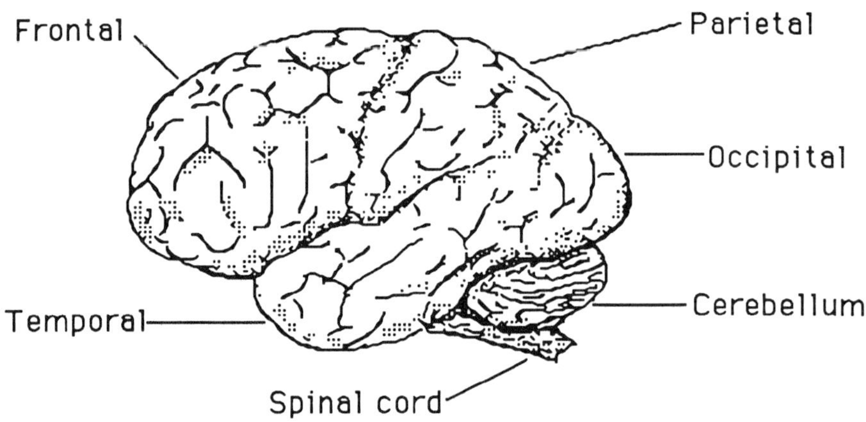

Fig. 11-1. Side view of the human brain

Various parts of the cerebral cortex are associated with very specific functions. The visual cortex is within the occipital lobe. The right side is associated with the left-hand field of vision of both eyes and the left visual cortex is associated with the right-hand field of vision of both eyes. Auditory cortex is part of the temporal lobe. The right deals mainly with sound received on the left and the left auditory cortex, on the whole, with sounds from the right. The olfactory cortex is in frontal lobe. It is an exception in that the right cortex deals largely with the right nostril and the left with the left nostril. The sensation of touching is associated with somatosensory cortex in parietal lobe just behind the division between the frontal and parietal lobes. There is a very specific correspondence between the various parts of the surface of the body and the regions of the somatosensory cortex. The motor cortex controls the movement of muscles. It is lying just in front of the division between the frontal and parietal lobes. There is a very specific correspondence between the various muscles of the body and the regions of the motor cortex.

The neocortex areas described above are termed primary cortex in Fig. 11-2. They are the ones most directly concerned with the input and output of the brain. Near to these primary regions are the secondary regions of the cerebral cortex, which are concerned with a more subtle and complex level of abstraction. The remaining regions of the cerebral cortex are referred to as tertiary (or the association cortex), where the most abstract and sophisticated activity of the brain is carried out. It is here that the information from various different sensory regions is integrated and analyzed in a very complex way, memories are laid down, general plans are conceived and evaluated, and believed that sensory input is translated into symbolic processes.

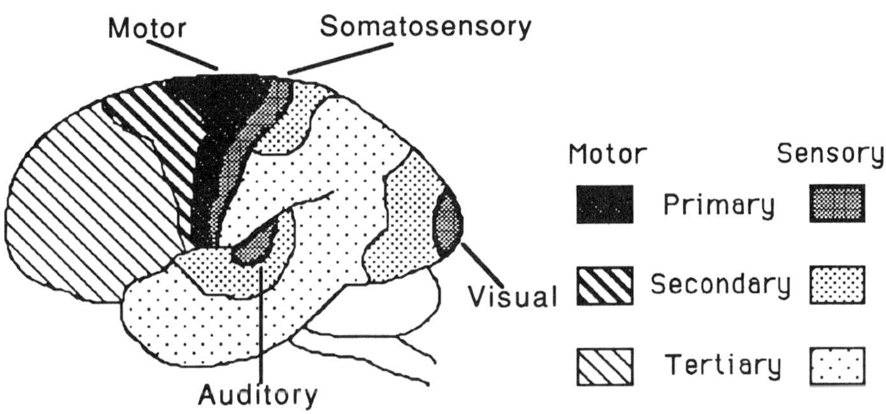

Fig. 11-2. Function hierarchy of cortex

In a general way, there is a movement of the brain's activity starting at the primary portions of the rear lobes, moving on to the secondary portions, as the input data are analyzed, and then on to the tertiary portions of the rear lobes as these data become fully comprehended (*e.g.*, as with the comprehension of speech at Wernicke's area). Then the processed information is carried to the frontal lobe, where general plans of action are formulated in its tertiary regions (*e.g.*, as with the formulation of speech in Broca's area). These general plans of action are translated into more specific conceptions about movements of the body in the secondary motor regions and, finally, the brain's activity moves on to the primary motor cortex where signals are sent on, ultimately, to various groups of muscles in the body (and often to several at once).

The above description is an oversimplified one, there also seems to be other subsidiary sensory and motor regions scattered about the cerebral cortex, however, it does give a flavor of brain's hierarchical organization.

Apart from the cerebrum, other relatively more important parts of the brain are shown in Fig. 11-3:

Cerebellum: it is responsible for sensory-motor coordination and the establishment of body equilibrium. It seems that, when one is learning a new skill, initially one must think through each action in detail, and the cerebrum is in control; but when the skill has been mastered, it is the cerebellum that takes over. Thinking about it seems to involve the re-introduction of cerebral control and, although a consequent flexibility of activity is introduced, the flowing and precise cerebellum action is lost.

Hippocampus: it plays a vital role in laying down long-term memories.

Corpus callosum: a region via which the right and left cerebral hemispheres communicate with one another.

Hypothalamus: it has only 0.3% of brain's weight yet involved in all aspect of brain behavior: feeding, sleeping, temperature regulation, endocrine function, movement. It is also considered to be the seat of emotion.

Thalamus: a processing center and relay station, it conveys many of the nerve inputs from the external world to the cerebral cortex.

Reticular formation: cell bodies and axons are mixed together here, it is responsible for the general state of alertness or awareness.

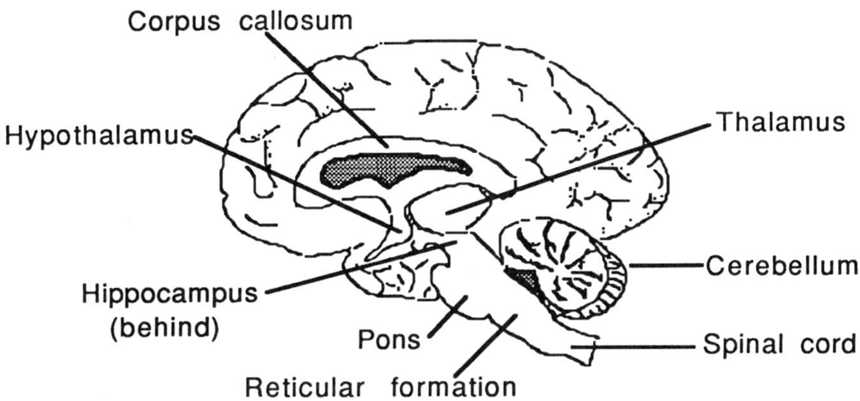

Fig. 11-3. A quick look at the brain.

Neuron Structure and Activity

Neurons are cells that are the integrating units of the nervous system, they are modifiable and most live as long as the person in which they are found. The neuron, shown in Fig. 11-4, consists of a cell body, processes called dendrites (from the Greek, meaning tree), a process called an axon (from the Greek, meaning axle). The dendrites collect information, which is then integrated at the axon hillock close to the cell body; a summary of the input received by the cell is then passed along the axon and through the end feet to other cells.

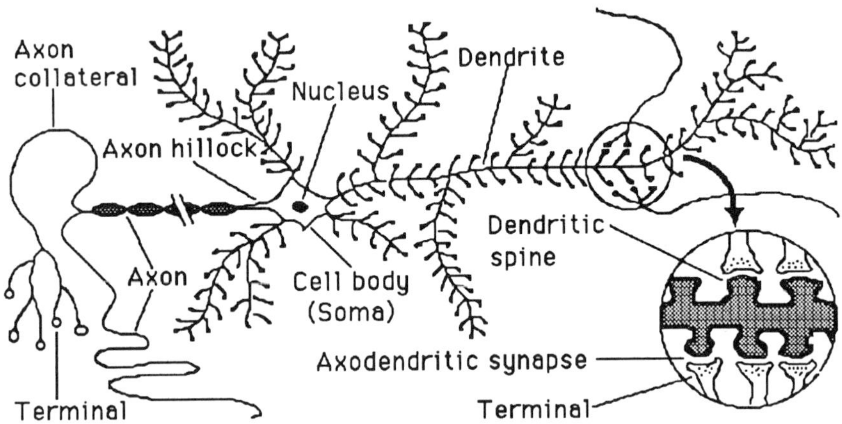

Fig. 11-4. Neuron

The dendrites are actually extensions of the cell body that allow the neuron to increase the area of surface on which it receives information from other cells. Some neurons have a few of them, others over 20. Dendrites vary from a few microns to millimeters in length and taper as they branch; some have rough projections called dendritic spines on which they receive terminals from other cells.

The axon originates in the cell body at a transition point called the axon hillock. Each cell has only one axon, which varies in length from a few microns to more than a meter in different cells. Most axons have branches called collaterals. At the end of the axon are little knobs called terminals.

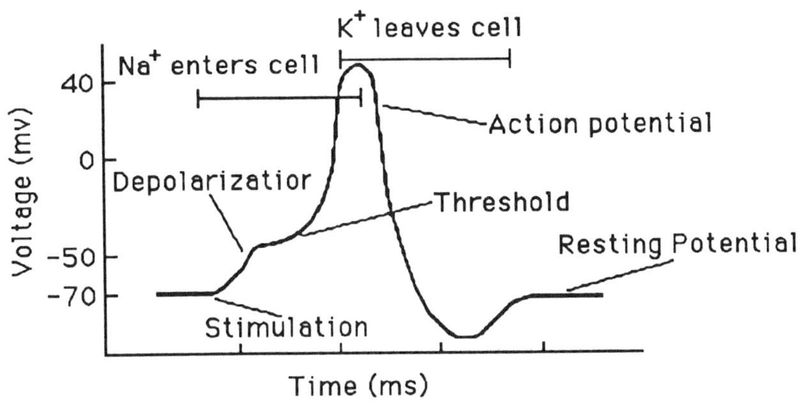

Fig. 11-5. Action potential

Due to the flow control of neuron membrane, there are 350 times as many organic ions (An^-) and 20 times as many potassium ions (K^+) on the inside of the cell membrane as on the outside, and 9 times as many chlorine ions (Cl^-) and 20 times as many sodium ions (Na^+) on the outside of the cell membrane as on the inside, resulting a voltage on the side of the cell in range of -70 *mv*. Stimulation, being electric current or chemical, of the membrane causes it to become more permeable to Na^+. When Na^+ enters the cell, the potential decreases. If the stimulation is sufficiently intense to cause the transmembrane voltage to depolarize to about -50 *mv*, its threshold, the membrane becomes completely permeable to Na^+, so Na^+ rushes into the cell until the voltage across the membrane falls through 0 *mv* and reverses to about +50 *mv*. The change in permeability is about half a millisecond, after which the membrane becomes permeable to K^+. The flow of K^+ out of cell restores and even overshoots the original resting potential.

Action potential occurs at one region of the nerve membrane serves as a stimulus to the adjacent portions of the membrane to lose their permeability and undergo a similar voltage change. This movement of action potential along the length of the axon is called a nerve impulse or firing.

A peculiar property of a neuron's behavior is that its threshold is stable, and every action potential, and hence nerve impulse, once it is triggered has an identical threshold and height. Once the nerve impulses are strong enough to pass the threshold they require no further stimulus; the neuron provides all the energy required to maintain them. These properties of the neuron's behavior are formulated in the all-or-none law: action potentials either occur or they do not; there is no in-between condition. If we want to strictly comply with the physical reality, it seems the binary activation function in Fig. 11-6, is more appropriate than the sigmoid function in artificial neural nets.

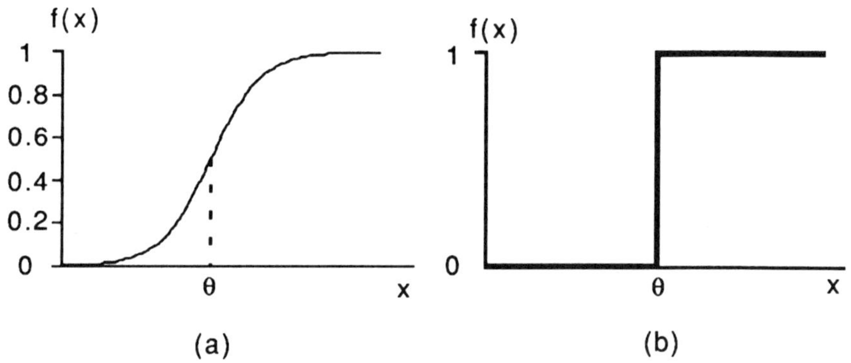

Fig. 11-6. Activation functions. (a) sigmoid. (b) binary.

Most axon terminals contain a number of tiny oval sacs called synaptic vesicles. When the impulse arrives, chemicals called neurotransmitter are released from vesicles into the synaptic space. The neurotransmitters then stimulate the postsynaptic neuron and develop a small potential of 1 to 3 *mv*, either positive (excitatory) or negative (inhibitory). The integration of many inputs by spatial and temporal summation determines whether the neuron will fire or not.

Dendrites have a membrane similar to the axons' and a similar resting potential, and they also undergo changes in potential when they are stimulated. But unlike the axon, dendrites do not produce action potentials. If a dendrite is stimulated, the voltage changes from the resting potential in proportion to the intensity of the stimulation; the change then spreads along the dendrite away from the point of stimulation, getting smaller with distance. These voltage changes undergone by dendrites are called graded potentials, which can occur as a decrease in transmembrane voltage (depolarization) or an increase (hyperpolarization), depending on the nature of the stimulation. Both intensity and location of the stimuli affect their influentialbility to the neuron. Graded potentials generated at different locations at different times all adds up at the axon hillock, if the threshold is reached, the axon will fire.

A neuron is not awfully fast. The speed at which the impulse travels along the axon or dendrites, varying from 1 to 100 meters per second. Usually neurons fire fewer than 100 times a second, but they can fire as frequently as 1000 times/s. It takes about 0.5 *ms* in getting the neurotransmitter across the synapse. In contrast, IC logic gates usually have switching time in *ns* range, also, signal travels a lot faster in wires. The strength of nervous system lies in the myriads of neuron inter-connections. It is estimated that there are 10^{11} neurons and 10^{15} links. Fig. 11-7 shows various possible connection configurations. All the neurons act asynchronously, based on the influences of the environments and the conditions of themselves. Each firing may have a deterministic course, the cumulative activity on the whole is unpredictable.

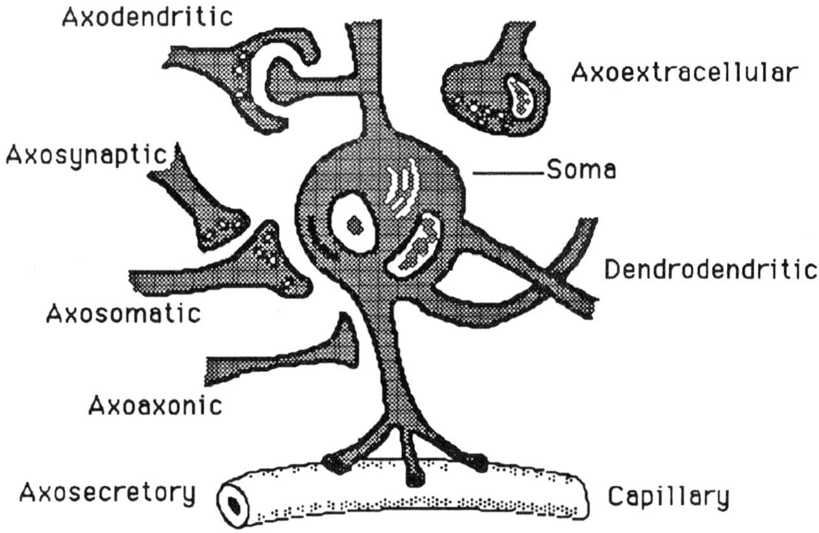

Fig. 11-7. Neuron connections.

General Principles of Brain Function

The functioning of brain is still far from well understood, however, there are some preliminary conclusions.

1. *Parallel* organization in the nervous system: Anatomical and electro-physiological evidences clearly indicate a parallel organization of the sensory, and probably the motor system. Thalamic-cortical connections are actually made up of parallel path ways with overlapping projections [Jones], as illustrated in Fig. 11-8.

2. *Hierarchical* organization: successive levels of interneurons permit successively more complex or more adaptive behavior [Gallistel]. Fig. 11-9 illustrates the pathway from sensory neurons to motor neurons. Through the connection of the interneurons each type of afferent stimulation can recruit a specific population of motor neurons and thus produce a specific movement. The same motor neurons can be used in different combinations to produce a wide variety of movements.

3. *Lateralization* of function: the brain is functionally and anatomically

asymmetrical. In most right-handed people the left hemisphere has a role in language and in controlling complex voluntary movements that is not shared by the right hemisphere, where the right hemisphere has a role in the control of visuospatial abilities not shared by the left hemisphere.

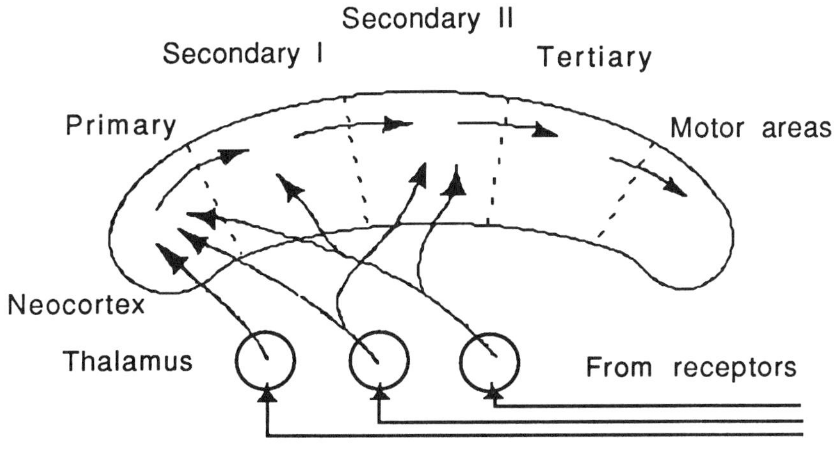

Fig. 11-8. Parallel brain.

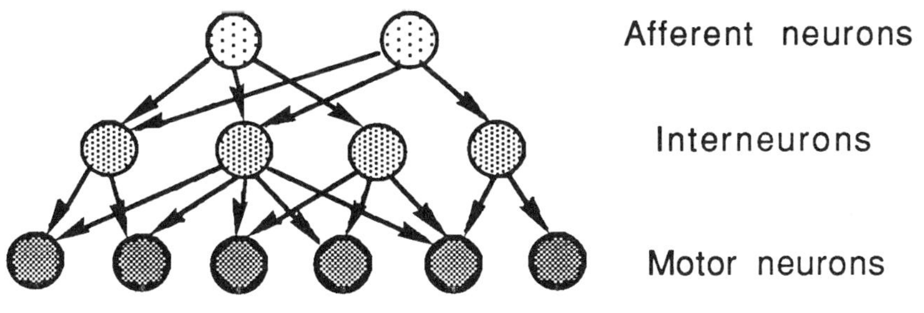

Fig. 11-9. Hierarchical brain.

It is relatively easier to study the nervous system's micro details, and to observe the human macro behavior, but it is hard to relate the psychological events with the physiological properties. One of the best attempts was made by Hebb. He hypothesized that "When an axon of cell *A* is near enough to excite a cell *B* and repeatedly or persistently takes part in firing it, some growth process or metabolic change takes place in one or both cells such that *A*'s efficiency, as one of the cells firing *B*, is increased." According to him, short-term memory is reverberation of the closed loops of the cell assembly; long-term memory requires structural synaptic changes to occur, and it needs a period of consolidation to make the change. Hebbian's law is often cited in artificial neural nets research.

11.2 ANALOG IMPLEMENTATIONS

The proposed artificial neural networks have a common principle where they follows the simple motion equation:

$$\frac{dU_i}{dt} = -\frac{\partial E(V_1,\ldots,V_n)}{\partial V_i} \qquad (11.1)$$

where $E(V_1,\ldots,V_n)$ is called the computational energy function. As long as the system composed of n processing elements (neurons) follows the motion equation, the value of E will monotonically decrease. In other words, the state of the system is allowed to converge to the local minimum.

Fig. 11-10 shows the analog neural network where each neuron performs the motion equation in Eq. (11.1). There are two types of neurons: excitory neurons and inhibitory neurons. The artificial neurons are interconnected by the artificial synaptic links which can be realized by resistors.

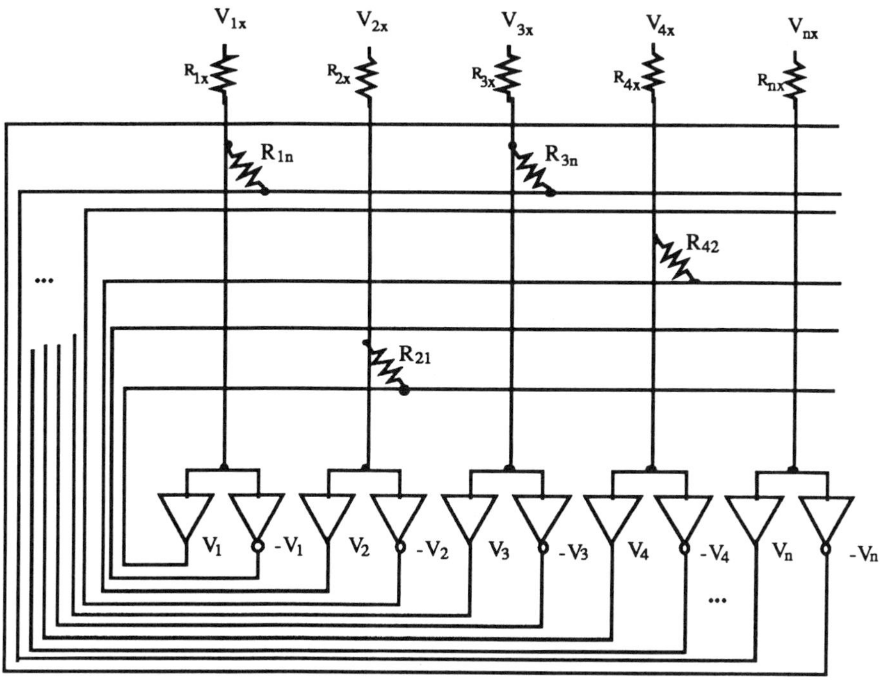

Fig. 11-10 A circuit diagram of the artificial neural network

In Fig. 11-11 the detailed neuron circuit is described where the operational amplifier on the first stage sums all of inputs which satisfies the following equation:

$$\frac{dU_i}{dt} = -R\left(\sum_{j=1}^{n^2} \frac{V_j}{R_{ij}} + \frac{V_{ix}}{R_{ix}}\right) \tag{11.2}$$

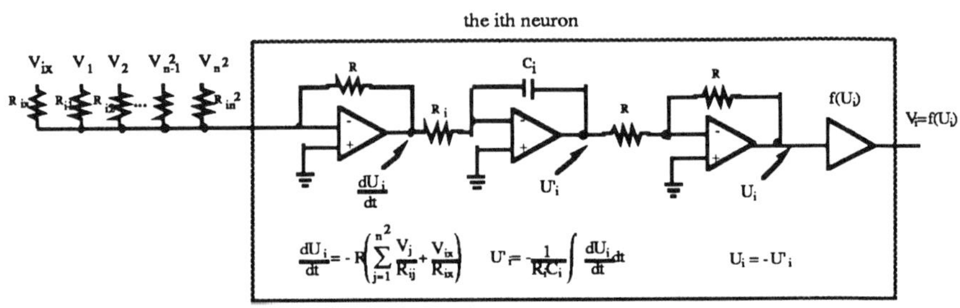

Fig. 11-11 An analog circuit of the *i*th McCulloch-Pitts neuron

For example, the derivatives of the four-color computational energy E with respect to the *p*th output as shown in Eq. (3.4) of Chapter 3 is given by:

$$\frac{dU_p}{dt} = - \sum_{q=1}^{n^2} G_{pq}V_q - I_p \tag{11.3}$$

where p and q are given by (4(X-1)+i) and (4(Y-1)+j) respectively. G_{pq} is given by:

$$G_{pq} = A\delta_{XY}(1-\delta_{ij}) + B\delta_{XY} + Cd_{XY}\delta_{ij}(1-\delta_{XY}) \tag{11.4}$$

$$\text{where } \delta_{rs} = \begin{cases} 1 & \text{if } r=s \\ 0 & \text{otherwise} \end{cases}$$

From Eq. (11.2) and (11.3), G_{pq} and I_p in Eq. (11.3) are determined by R/R_{ij} and RV_{ix}/R_{ij} respectively.

The second operational amplifier in Fig. 11-11 performs integration of dU_i/dt which is given by the following equation:

$$U'_i = -\frac{1}{R_iC_i}\int \left(\frac{dU_i}{dt}\right)dt \tag{11.5}$$

The third operational amplifier generates U_i, $U_i = -U'_i$. The last component must perform a nondecreasing function such as the sigmoid or the binary function.

The hill-climbing term in the Newton equation can be easily implemented by the circuit described in Fig. 11-12 when binary neurons are used. If and only if $V_{x1}=V_{x2}=V_{x3}=V_{x4}=0$, $-V_x$ will be given to the input of the ith color of the X state through the analog switch and the resistor R_x , otherwise the circuit of the resistor will be cut off.

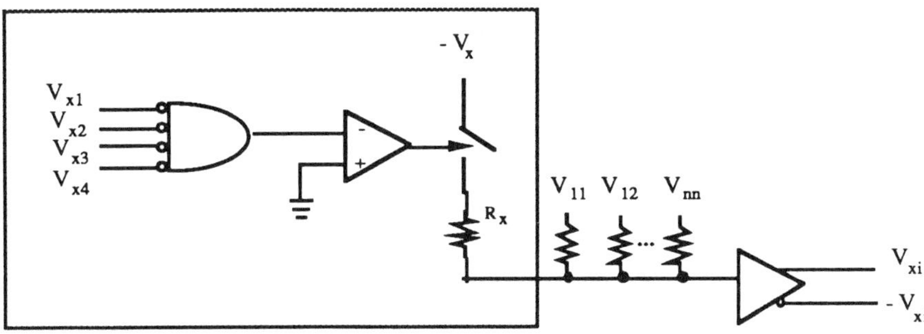

Fig. 11-12 The circuit of the *i*th neuron in the *X*th state for the hill-climbing term

11.3 DIGITAL IMPLEMENTATIONS

The output of the *i*th hysteresis McCulloch-Pitts binary neuron V_i is given by :

$$V_i = \begin{cases} 1 & \text{if } U_i \geq \text{UTP (upper trip point)} \\ 0 & \text{if } U_i \leq \text{LTP (lower trip point)} \\ \text{unchange} & \text{if LTP} < U_i < \text{UTP}, \end{cases} \tag{11.6}$$

where U_i is the input of the *i*th neuron. The output of the hysteresis neuron depends upon the present value of the input and the past values of the output.

Layout design of the hysteresis McCulloch-Pitts binary neuron

It is easier to implement neural networks using digital circuits than using analog circuits because the digital approach is more tolerant of unavoidable variations.

Furthermore, the CMOS digital VLSI technology is one of the most successful technologies today. VLSI layout design of the hysteresis McCulloch-Pitts binary neuron using the CMOS digital VLSI technology is presented.

Our simulation results using the hysteresis McCulloch-Pitts binary neuron model give the following facts:

1) The computation resolution with the range from -15 to 10 in integer number is sufficient for the correct motion of each neuron.

2) The maximum hysteresis range is 10. If the hysteresis range is too big, the convergence speed of the system was found to be very low.

Based on these facts, we set the following architecture for CMOS VLSI design of the hysteresis McCulloch-Pitts binary neurons. To realize the hysteresis property, we design a sequential circuit using flip-flop programming. The fixed point 2's complement arithmetic is used for calculation of the input which requires additions and subtractions only. To express the value of the input, we use a 6-bit with 2's complement arithmetic which is sufficient to express the integer range from -15 to 10. For simplicity, LTP is set at 0 and UTP can be reset between 0 and 7. Thus, the maximum hysteresis range can be expressed with 3-bit.

Fig. 11-13 shows the state transition of the hysteresis McCulloch-Pitts binary neuron. Here, two circles represent two different states, where the output of the neuron is 0 or 1 respectively. Each arrow indicates the state transition. The numbers attached to each arrow represent the range of the input.

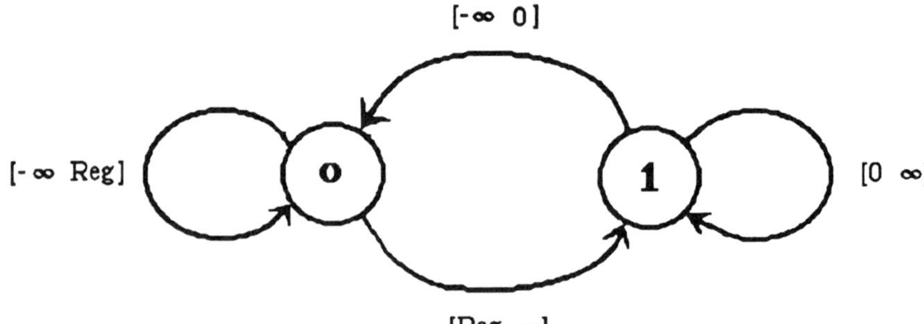

Fig. 11-13 State transitions of the hysteresis McCulloch-Pitts binary neuron

Fig. 11-14 shows the digital circuit diagram of the hysteresis McCulloch-Pitts binary neuron. The 5-bit input U(U0,U1,⋯,U5) is compared at the 5-bit comparator with the specific value which will be equal to either LTP or UTP. A multiplexer realized by three NOR gates transmits the value of LTP or the value of UTP according to the previous state of the neuron itself which is labeled as "V" and is stored in a leftmost flip-flop in Fig. 11-14. The state of a neuron is updated synchronously with a clock signal labeled "CK2". The 3-bit value of UTP is stored in the 3-bit shift register through the serial input line named "Hys". For this purpose, another clock signal labeled "CK1" is used. To initialize the whole circuit, a line named "$\overline{\text{CLR}}$" is prepared. This line is connected with a reset circuit including the power-on reset circuit and the manual reset circuit which is fabricated out of the VLSI chip.

The proposed circuit of the hysteresis McCulloch-Pitts binary neuron using the CMOS VLSI technology is simulated using a logic simulator on a Macintosh SE/30. Fig. 11-15a shows the timing diagram of our simulation result. At the beginning of our simulation, one negative pulse is given to $\overline{\text{CLR}}$ in order to reset the flip-flops for the initial state. Then, after three clock cycles of CK1, 3-bit value as the maximum hysteresis range is loaded into the 3-bit shift register. To simulate the hysteresis function, twelve state changes, each state is denoted by A,B,⋯,M in Fig. 11-15b, are given to the circuit. The correct motion of the output V of the hysteresis McCulloch-Pitts binary neuron is confirmed.

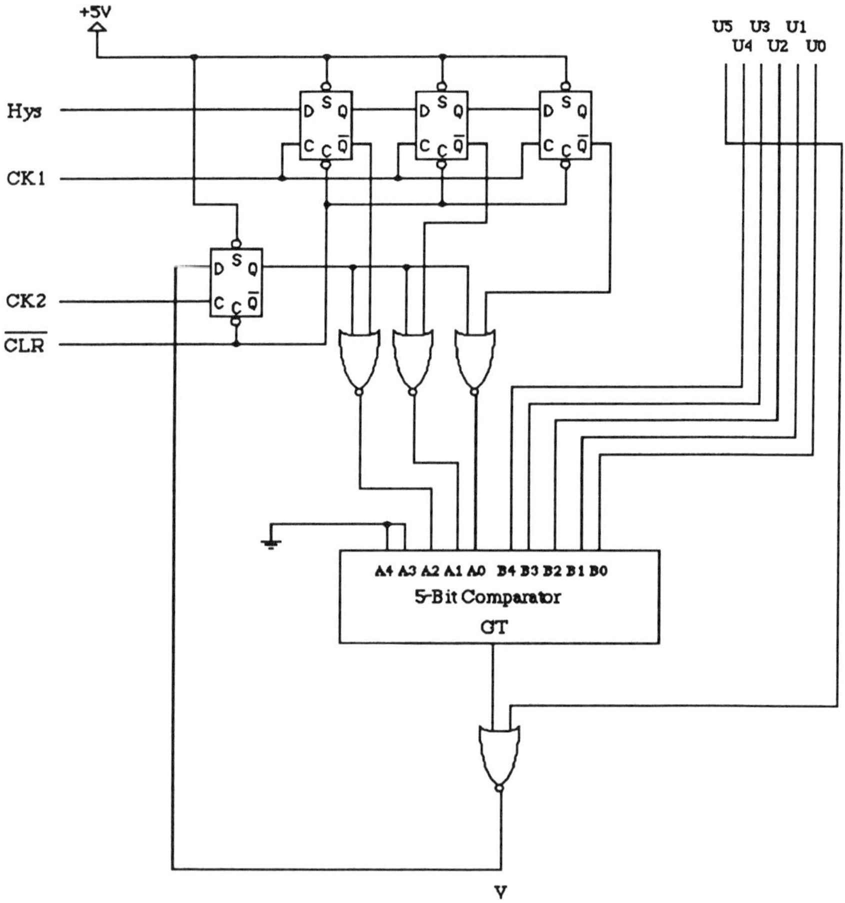

Fig. 11-14 Digital hysteresis McCulloch-Pitts neuron

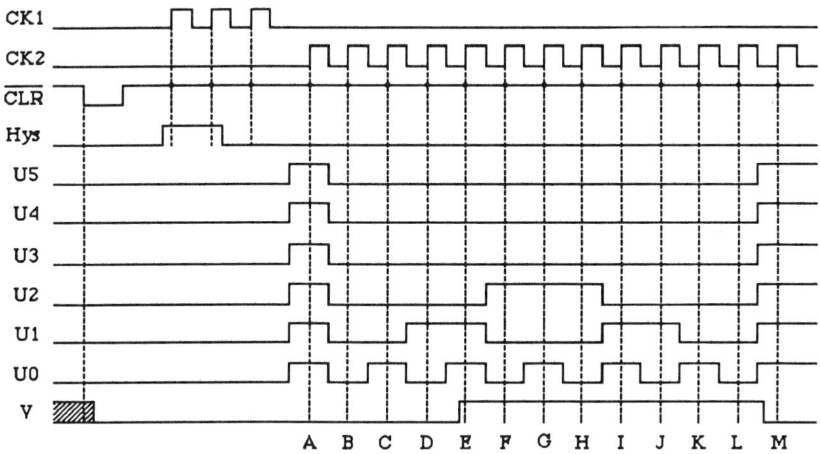

Fig. 11-15 (a) Timing diagram of the simulation result

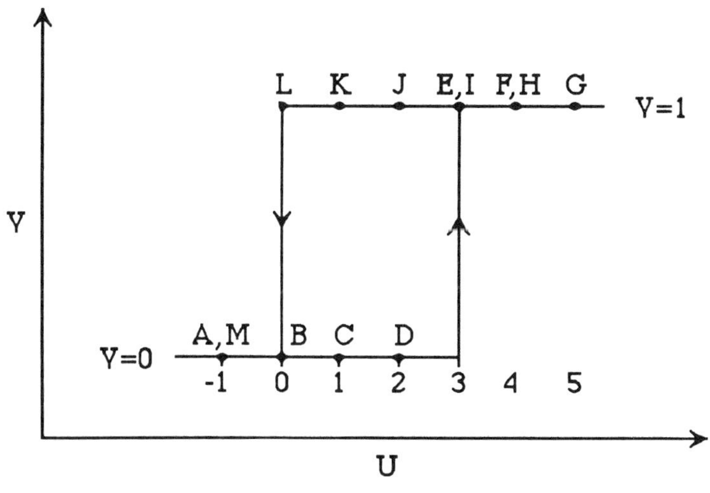

Fig. 11-15 (b) Thirteen states of the simulated points

11.4 REFERENCES

Foo S. Y., L. R. Anderson L. R., Takefuji Y., (July 1990), Analog components for the VLSI of artificial neural networks, IEEE Circuits and Devices, pp.18-26.

Jones, E.G., 1981, Anatomy of cerebral cortex: Columnar input-output organization. In F. O. Schmitt, F. G. Worden, G. Adelman, and S. G. Dennis, eds., *The Organization of the Cerebral Cortex.* Cambridge, Mass: MIT Press.

Kolb, Bryan and Ian Q. Whishaw: Fundamentals of Human Neuropsychology, 2nd Ed. New York: W. H. Freeman and Company, 1985.

Kurokawa T., Lee K. C., Cho Y. B., and Takefuji Y., 1990, CMOS layout design of the hysteresis McCulloch-Pitts neuron, Electronics Letters, 26, 25, pp.2093-2095.

Lashley, K.S.: Functional determinants of Cerebral Localization. In F. A. Beach, D. O. Hebb, C. T. Morgan, and H. W. Nissen, eds., *The Neuropsychology of Lashley.* New York: McGraw-Hill, 1960.

Luria, A.R.: The Working Brain. Harmondsworth: Penguin, 1973.

MacGregor, Ronald J.: Neural and Brain Modeling. San Diego, CA: Academic Press, Inc., 1987.

Pribram, K.H.: Languages of the Brain. Englewood Cliffs, N.J.: Prentice-Hall, 1971.

Skarda, Christine A. and Walter J. Freeman: How Brains Make Chaos in Order to Make Sense of the Takefuji Y., and K. C. Lee, (March 1991), Artificial neural networks for four-coloring problems and k-colorability problems, IEEE Trans. on Circuits and Systems, 38, 3, pp.326-333.

11.5 EXERCISES

1. In order to solve the four-coloring map problem in Fig. 11-16, complete the circuit in Fig. 11-17.

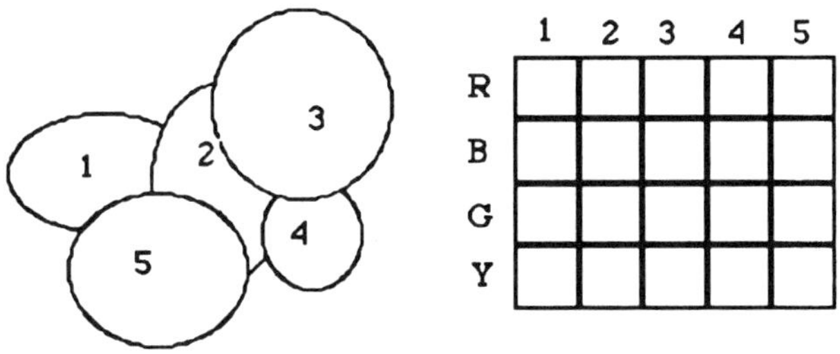

Fig. 11-16 Five-country map and neural representation

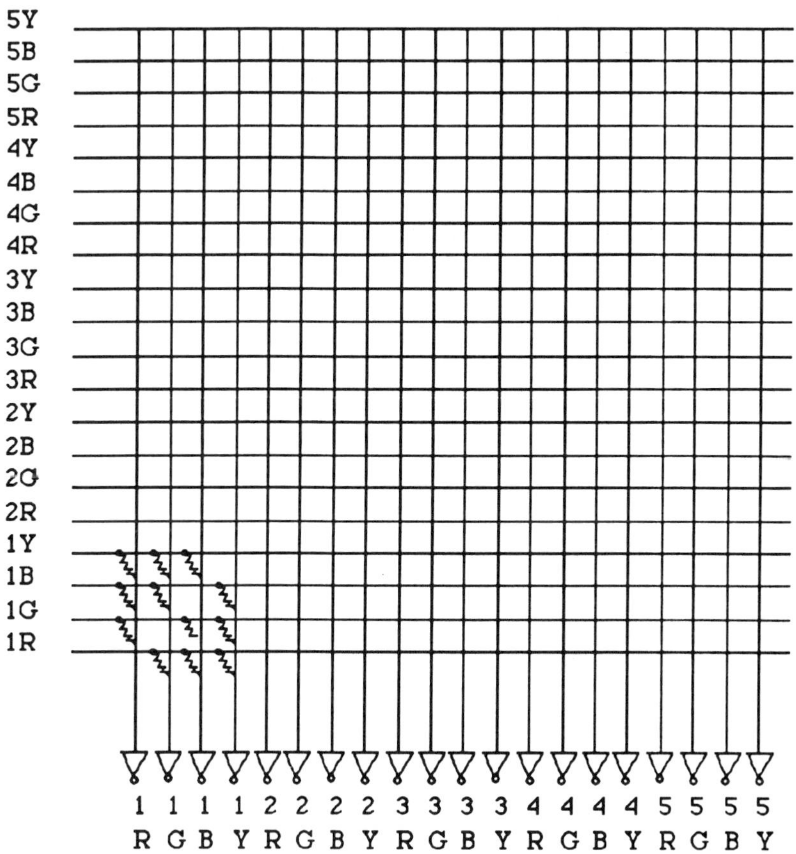

Fig. 11-17 Four-coloring circuit for the five-country map

2. Show how to construct the 12 x 12 knight's tour using the following solutions.

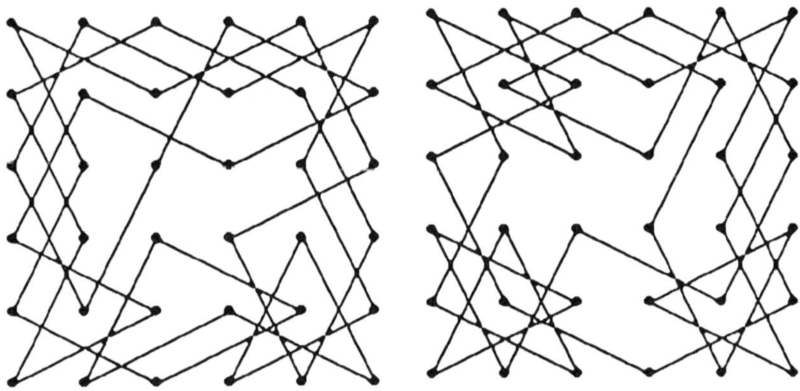

Fig. 11-18 Solutions of the 6 x 6 knight's tour problem

3. Design the circuit for the 6 x 6 knight tour problem.

4. Survey research on silicon neural network implementations and optical implementations.

Chapter 12

MATHEMATICAL BACKGROUND

In this Chapter a mathematical background for computing artificial neural network models and solving optimization problems is given and four neuron models are introduced to show computational mathematics. Although it is still widely believed that the decay term in the motion equation is absolutely essential, Section 12.2 shows that the long-time common belief on the decay term is denied and the removal of the decay term from the motion equation is highly recommended to guarantee the local minimum convergence. The local minimum convergence of the continuous system using the motion equation without the decay term is given in Section 12.3. The local minimum convergence of the discrete system is given in Section 12.4. McCulloch-Pitts neural network convergence theorem and proof are introduced in Section 12.5. Hysteresis McCulloch-Pitts neural network convergence theorem and proof are introduced in Section 12.6. A new neuron model called maximum neuron is introduced and the local minimum convergece is given. Finally other neuron models are discussed.

12.1 INTRODUCTION AND FOUR NEURON MODELS

An artificial neuron model has input and output where the relationship of the input/output function is given by:

$$V_i=f_{model}(U_i). \qquad (12.1)$$

Note that V_i and U_i is the output and the input of the *i*th neuron respectively. f_{model} is called input/output function.

McCulloch and Pitts introduced the first mathematical neuron model in 1943 (McCulloch and Pitts 1943). The input/output function of the McCulloch-Pitts neuron model is given by:

$$V_i=f_{MP}(U_i)=1 \text{ if } U_i>0 \text{ and } 0 \text{ otherwise.} \qquad (12.2)$$

The characteristics of the McCulloch-Pitts function is nondifferentiable, nonlinear, nondecreasing, and has two binary states.

Hopfield introduced the sigmoid neuron model (Hopfield 1982) where the input/output function is given by:

$$V_i=f_{sigmoid}(U_i)=\frac{1}{2}(\tanh(\lambda_0 U_i)+1) \qquad (12.3)$$

where λ_0 is called gain parameter. The characteristics of the sigmoid function are differentiable (continuous), nonlinear, nondecreasing, and has infinite states. The convergence of the McCulloch-Pitts neural network is relatively faster than that of the sigmoid neural network because the McCulloch-Pitts neuron has fewer states.

However the McCulloch-Pitts neural network sometimes shows the oscillatory behavior in neural dynamics. In order to suppress the undesirable oscillatory behavior, hysteresis is embedded in the McCulloch-Pitts neuron model which is called hysteresis McCulloch-Pitts neuron model. The input/output function of the hysteresis McCulloch-Pitts neuron model is given by:

$V_i=1$ if $U_i>$UTP (Upper Trip Point)

$=0$ if $U_i<$LTP (Lower Trip Point)

unchanged otherwise (12.4)

where UTP is always larger than LTP. Because of suppressing the oscillatory behavior it shortens the convergence time consequently. Unfortunately there is no theorem how to determine optimum UTP and LTP respectively. However the empirical study on values of UTP and LTP has been intensively investigated in solving crossbar switch scheduling problems in Chapter 2.

Takefuji introduced the maximum neuron model where the McCulloch-Pitts neuron model is modified and the winner-take-all function is embedded in the model. The maximum neural network is composed of M clusters where each cluster consists of n neurons. One and only one neuron among n neurons with the maximum input per cluster is encouraged to fire in the maximum neural network. The input/output function of the *i*th maximum neuron in the *m*th cluster is given by:

$$V_{m\,i} = 1 \text{ if } U_{m\,i} = \max\{ U_{m\,1}, .., U_{m\,n} \} \text{ and } U_{m\,i} \geq U_{m\,j} \text{ for } i > j,$$
$$= 0 \text{ otherwise.} \qquad (12.5)$$

where max is a function to choose the maximum value. In the McCulloch-Pitts neural network with/without hysteresis or in the sigmoid neural network it is not always guaranteed that the converged state is equivalent to an acceptable or satisfactory solution. The acceptable solution means that it satisfies the required constraints, however it may or may not be the best solution. In other words, the converged state is not always a satisfactory solution in the McCulloch-Pitts neural network or in the sigmoid neural network. However in the maximum neural network it is always guaranteed to generate satisfactory solutions. Because the maximum neural network always keeps one and only one neuron to fire per cluster. The advantages of the maximum neural network are summarized by: 1) every local minimum state is equivalent to the feasible (acceptable and satisfactory) solution while the other existing neural networks cannot guarantee it; 2) turing coefficients parameters in the motion equation is not required while the other existing neural network must suffer from it; and 3) the condition of the equilibrium state is clearly defined while the other existing neural network did not provide the clear definition.

12.2 WHY IS THE DECAY TERM HARMFUL?

The goal of using the artificial neural network is to minimize a fabricated computational energy function E where it is given by considering the necessary and sufficient constraints and the cost function. The computational energy function E is given by: $E=E(V_1, V_2,...,V_n)$. Before showing why the artificial neural network can minimize the fabricated function $E(V_1, V_2,...,V_n)$, "why is the decay term is harmful?" is shown although it has been widely believed that the decay term is absolutely essential. The motion equation of the *i*th neuron with the decay term is generaly given by:

$$\frac{dU_i}{dt} = -\frac{U_i}{\tau} - \frac{\partial E}{\partial V_i} \tag{12.6}$$

where "$\frac{U_i}{\tau}$" in Eq.(12.6) is the controversial decay term. Note that τ is a constant parameter. In order to mathematically prove the use of the decay term in the motion equation harmful, the conditions to increase the fabricated energy function E instead of decreasing it are given in Theorem 1.

Theorem 1 :

The use of the decay term in Eq.(12.6) increases the computational energy E when $\left|\sum_i \frac{dV_i}{dt}\frac{U_i}{\tau}\right| > \left|\sum_i \left(\frac{dV_i}{dU_i}\right)\left(\frac{dU_i}{dt}\right)^2\right|$ and if either ($U_i > 0$ and $\frac{dV_i}{dt} < 0$) or ($U_i < 0$ and $\frac{dV_i}{dt} > 0$) is satisfied.

Proof:

Consider the derivatives of the computational energy E with respect to time t.

$$\frac{dE}{dt}=\sum_i \frac{dV_i}{dt}\frac{\partial E}{\partial V_i} = \sum_i \frac{dV_i}{dt}\left(-\frac{U_i}{\tau} - \frac{dU_i}{dt}\right)$$ where $\frac{\partial E}{\partial V_i}$ is replaced by

$(-\frac{U_i}{\tau} - \frac{dU_i}{dt})$ from Eq. (12.6)

$$= -\sum_i \frac{dV_i}{dt}\frac{U_i}{\tau} - \sum_i \frac{dV_i}{dt}\frac{dU_i}{dt} = -\sum_i \frac{dV_i}{dt}\frac{U_i}{\tau} - \sum_i \left(\frac{dU_i}{dt}\frac{dV_i}{dU_i}\right)\left(\frac{dU_i}{dt}\right)$$

$$= -\sum_i \frac{dV_i}{dt}\frac{U_i}{\tau} - \sum_i \left(\frac{dV_i}{dU_i}\right)\left(\frac{dU_i}{dt}\right)^2$$

The first term $-\sum_i \frac{dV_i}{dt}\frac{U_i}{\tau}$ can be positive, negative, or zero.

The second term $-\sum_i \left(\frac{dV_i}{dU_i}\right)\left(\frac{dU_i}{dt}\right)^2$ is always negative or zero, because the output $V_i=f(U_i)$ is a nondecreasing function. The following condition can be true:

$-\sum_i \frac{dV_i}{dt}\frac{U_i}{\tau} - \sum_i \left(\frac{dV_i}{dU_i}\right)\left(\frac{dU_i}{dt}\right)^2 > 0$ when $\left|\sum_i \frac{dV_i}{dt}\frac{U_i}{\tau}\right| > \left|\sum_i \left(\frac{dV_i}{dU_i}\right)\left(\frac{dU_i}{dt}\right)^2\right|$ and if one of the following condition is satisfied: ($U_i > 0$ and $\frac{dV_i}{dt} < 0$) or ($U_i < 0$ and $\frac{dV_i}{dt} > 0$). Under such a condition the derivatives of E with respect to time t must be positive; $\frac{dE}{dt} > 0$.

Therefore, the decay term increases the energy function under such conditions, which contradicts the conventional convergence theorem. **Q.E.D.**

The harmfulness of using the decay term in the motion equation can be easily tested by empirical simulation with varying the value of τ. In order to satisfy the convergence to the local minimum, we must eliminate the decay term from the motion equation. Theorem 2 states that the computational energy function E monotonically decreases regardless of the condition of the symmetry and diagonal constraints in the conductance matrix as long as the neurons obey a nondecreasing function and the motion equation of the *i*th neuron is given by:

$$\frac{dU_i}{dt} = -\frac{\partial E}{\partial V_i}. \qquad (12.7)$$

12.3 ANALOG NEURAL NETWORK CONVERGENCE

Without the decay term in the motion equation in Eq. (12.7) the derivatives of the computational energy $E=E(V_1, V_2,...,V_n)$ with respect to time t must monotonically decrease.

Theorem 2:

$\frac{dE}{dt} \leq 0$ is satisfied under two conditions such as $\frac{dU_i}{dt} = -\frac{\partial E}{\partial V_i}$ and $V_i = f(U_i)$

where $f(U_i)$ is a nondecreasing function.

Proof :

$$\frac{dE}{dt} = \sum_i \frac{dU_i}{dt}\frac{dV_i}{dU_i}\frac{\partial E}{\partial V_i}$$

$$= -\sum_i \left(\frac{dU_i}{dt}\right)^2 \frac{dV_i}{dU_i} \quad \text{where } \frac{\partial E}{\partial V_i} \text{ is replaced by } -\frac{dU_i}{dt} \text{ (condition 1)}$$

$$\leq 0 \quad \text{where } \frac{dV_i}{dU_i} > 0 \text{ (condition 2)} \quad \textbf{Q.E.D.}$$

Theorem 2 guarantees the convergence of the continuous system. However, the system is usually simulated on the digital computer. There always exist errors between the real values and the quantized one. For example, the discrete sigmoid function must be used in the digital computer instead of the continuous one. In Theorem 3, the convergence theory of the discrete sigmoid neural network is given.

12.4 DISCRETE SIGMOID NEURAL NETWORK CONVERGENCE

Theorem 3:

$\frac{\Delta E}{\Delta t} \leq 0$ is satisfied under two conditions such as $\frac{\Delta U_i}{\Delta t} = -\frac{\Delta E}{\Delta V_i}$ and $V_i = f(U_i)$

where $f(U_i)$ is a discrete nondecreasing function.

Proof :

$$\frac{\Delta E}{\Delta t} = \sum_i \frac{\Delta U_i}{\Delta t}\frac{\Delta V_i}{\Delta U_i}\frac{\Delta E}{\Delta V_i}$$

$$= -\sum_i \left(\frac{\Delta U_i}{\Delta t}\right)^2 \frac{\Delta V_i}{\Delta U_i} \quad \text{where } \frac{\Delta E}{\Delta V_i} \text{ is replaced by } -\frac{\Delta U_i}{\Delta t} \text{ (condition 1)}$$

$$\leq 0 \quad \text{where } \frac{\Delta V_i}{\Delta U_i} > 0 \text{ (condition 2)}$$

Let $\frac{\Delta U_i}{\Delta t}$ be $\frac{U_i(t+\Delta t)-U_i(t)}{\Delta t}$. Let $\frac{\Delta V_i}{\Delta U_i}$ be $\frac{V_i(t+\Delta t)-V_i(t)}{U_i(t+\Delta t)-U_i(t)}$. It is necessary and sufficient to consider the following five cases as shown in Fig. 12-1 :

1) $U_i(t+Dt) > U_i(t)$, $V_i(t+Dt) = V_i(t)$
2) $U_i(t+Dt) > U_i(t)$, $V_i(t+Dt) > V_i(t)$
3) $U_i(t+Dt) < U_i(t)$, $V_i(t+Dt) < V_i(t)$
4) $U_i(t+Dt) < U_i(t)$, $V_i(t+Dt) = V_i(t)$
5) $U_i(t+Dt) = U_i(t)$, $V_i(t+Dt) = V_i(t)$

case 1) : $\frac{\Delta V_i}{\Delta U_i} = 0 \Rightarrow \frac{\Delta E}{\Delta t} = 0$

case 2) : $\frac{\Delta U_i}{\Delta t} > 0$ and $\frac{\Delta V_i}{\Delta U_i} > 0 \Rightarrow \frac{\Delta E}{\Delta t} < 0$

case 3) : $\frac{\Delta U_i}{\Delta t} < 0$ and $\frac{\Delta V_i}{\Delta U_i} < 0 \Rightarrow \frac{\Delta E}{\Delta t} < 0$

case 4) : $\frac{\Delta V_i}{\Delta U_i} = 0 \Rightarrow \frac{\Delta E}{\Delta t} = 0$

case 5) : $\frac{\Delta U_i}{\Delta t} = 0 \Rightarrow \frac{\Delta E}{\Delta t} = 0$

It can be concluded that the energy function E monotonically decreases as long as the motion equation of the neurons is given by $\frac{\Delta U_i}{\Delta t} = -\frac{\Delta E}{\Delta V_i}$ and $V_i = f(U_i)$ where $f(U_i)$ is a discrete nondecreasing function. **Q.E.D.**

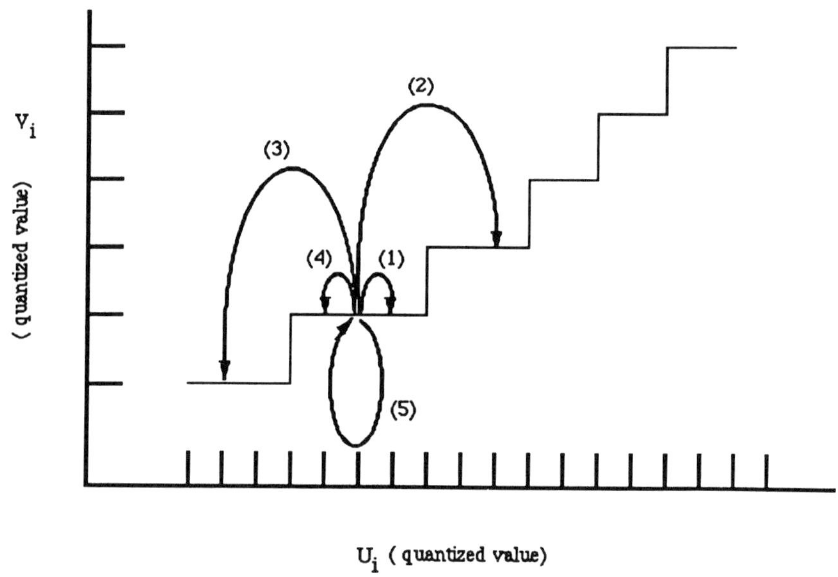

Fig. 12-1 Five possible state transitions in the discrete sigmoid neuron

12.5 MCCULLOCH-PITTS NEURAL NETWORK CONVERGENCE

In Theorem 4 , the convergence theory of the McCulloch-Pitts neural network is given.

Theorem 4:

$\frac{\Delta E}{\Delta t} \leq 0$ is satisfied under two conditions such as $\frac{\Delta U_i}{\Delta t} = -\frac{\Delta E}{\Delta V_i}$ and $V_i = f(U_i)$ where $f(U_i)$ is a binary function: $f(U_i) = 1$ if $U_i \geq 0$, 0 otherwise.

Proof:

Consider the derivatives of the computational energy E with respect to time t.

$$\frac{\Delta E}{\Delta t} = \sum_i \frac{\Delta U_i}{\Delta t}\frac{\Delta V_i}{\Delta U_i}\frac{\Delta E}{\Delta V_i}$$

$$= - \sum_i \left(\frac{\Delta U_i}{\Delta t}\right)^2 \frac{\Delta V_i}{\Delta U_i} \quad \text{where } \frac{\Delta E}{\Delta V_i} \text{ is replaced by } -\frac{\Delta U_i}{\Delta t} \text{ (condition 1)}$$

$$\leq 0 \quad \text{where } \frac{\Delta V_i}{\Delta U_i} > 0 \text{ (condition 2)}$$

Let $\frac{\Delta U_i}{\Delta t}$ be $\frac{U_i(t+\Delta t)-U_i(t)}{\Delta t}$. Let $\frac{\Delta V_i}{\Delta U_i}$ be $\frac{V_i(t+\Delta t)-V_i(t)}{U_i(t+\Delta t)-U_i(t)}$. It is necessary and sufficient to consider the following seven cases as shown in Fig. 10-2:

1) $U_i(t+Dt) > U_i(t)$, $U_i(t+Dt) < 0$, and $U_i(t) < 0$

2) $U_i(t+Dt) > U_i(t)$, $U_i(t+Dt) \geq 0$, and $U_i(t) < 0$

3) $U_i(t+Dt) > U_i(t)$, $U_i(t+Dt) > 0$, and $U_i(t) \geq 0$

4) $U_i(t+Dt) < U_i(t)$, $U_i(t+Dt) \geq 0$, and $U_i(t) > 0$

5) $U_i(t+Dt) < U_i(t)$, $U_i(t+Dt) < 0$, and $U_i(t) \geq 0$

6) $U_i(t+Dt) < U_i(t)$, $U_i(t+Dt) < 0$, and $U_i(t) < 0$

7) $U_i(t+Dt) = U_i(t)$

If the condition 7) is satisfied, then $\frac{\Delta U_i}{\Delta t}=0$ must be zero so that $\frac{\Delta E}{\Delta t}=0$. If 1), 3), 4), or 6) is satisfied, then $\frac{\Delta V_i}{\Delta U_i}$ must be zero, because of $\frac{\Delta V_i}{\Delta U_i} = \frac{V_i(t+\Delta t) - V_i(t)}{U_i(t+\Delta t) - U_i(t)} = \frac{0}{\text{nonzero number}}$, so that $\frac{\Delta E}{\Delta t}=0$. If 2) condition is satisfied, then $\frac{\Delta V_i}{\Delta U_i}$ must be positive, because of $\frac{\Delta V_i}{\Delta U_i} = \frac{V_i(t+\Delta t) - V_i(t)}{U_i(t+\Delta t) - U_i(t)}$

$= \frac{1}{\text{positive number}}$, so that $\frac{\Delta E}{\Delta t}<0$. If 5) condition is satisfied, then $\frac{\Delta V_i}{\Delta U_i}$ must be also positive, because of $\frac{\Delta V_i}{\Delta U_i} = \frac{V_i(t+\Delta t) - V_i(t)}{U_i(t+\Delta t) - U_i(t)} = \frac{-1}{\text{negative number}}$, so that $\frac{\Delta E}{\Delta t}<0$.

We can conclude that the energy function E monotonically decreases as long as the motion equation of the binary neurons is given by $\frac{\Delta U_i}{\Delta t} = -\frac{\Delta E}{\Delta V_i}$. **Q.E.D.**

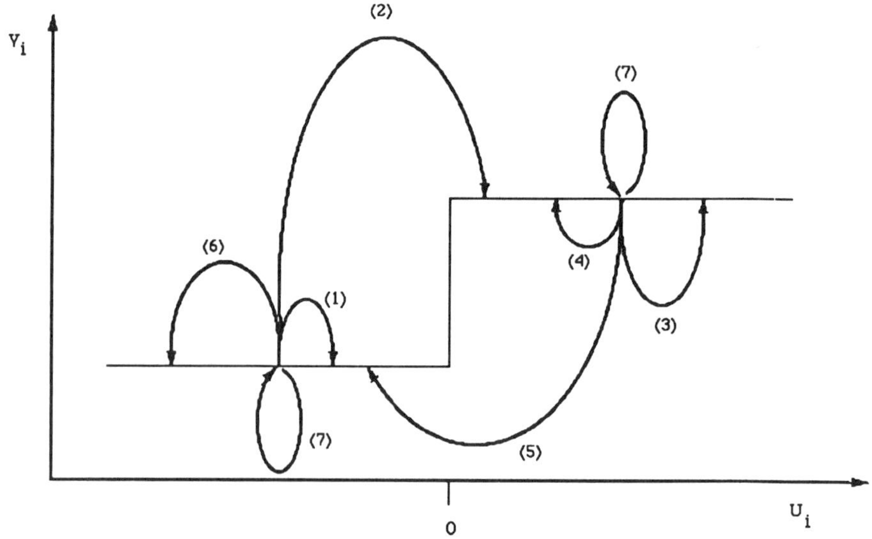

Fig. 12-2 Seven possible state transitions in the McCulloch-Pitts neuron

The McCulloch-Pitts neural network provides faster convergence than the sigmoid neural network. The McCulloch-Pitts neural network sometimes introduces the undesirable oscillatory behavior. However, the hysteresis property of the hysteresis McCulloch-Pitts neural networks suppresses the oscillatory behavior.

12.6 HYSTERESIS MCCULLOCH-PITTS NEURAL NETWORK

The output of the *i*th hysteresis McCulloch-Pitts binary neuron V_i is given by : V_i =1 if U_i ≥UTP (upper trip point), 0 if U_i ≤LTP (lower trip point), and V_i is unchanged otherwise where U_i is the input of the *i*th neuron. The output at any moment does not depend only upon the present value of the input but also upon the past values. In Theorem 5, the convergence theory of the hysteresis McCulloch-Pitts neural network is given.

Theorem 5 :

It is satisfied $\frac{\Delta E}{\Delta t} \leq 0$ under two conditions such as $\frac{\Delta U_i}{\Delta t} = -\frac{\Delta E}{\Delta V_i}$ and $V_i = f(U_i)$ where $f(U_i)$ is a hysteresis binary function.

Proof :

Consider the derivatives of the computational energy E with respect to time t.

$$\frac{\Delta E}{\Delta t} = \sum_i \frac{\Delta V_i}{\Delta t}\frac{\Delta E}{\Delta V_i} = \sum_i \frac{\Delta V_i}{\Delta t}\left(-\frac{\Delta U_i}{\Delta t}\right) \text{ where } \frac{\Delta E}{\Delta V_i} \text{ is replaced by } \left(-\frac{\Delta U_i}{\Delta t}\right)$$

$$= -\sum_i \left(\frac{\Delta U_i}{\Delta t}\frac{\Delta V_i}{\Delta U_i}\right)\left(\frac{\Delta U_i}{\Delta t}\right) = -\sum_i \left(\frac{\Delta V_i}{\Delta U_i}\right)\left(\frac{\Delta U_i}{\Delta t}\right)^2$$

Let $\frac{\Delta U_i}{\Delta t}$ be $\frac{U_i(t+\Delta t)-U_i(t)}{\Delta t}$. Let $\frac{\Delta V_i}{\Delta U_i}$ be $\frac{V_i(t+\Delta t)-V_i(t)}{U_i(t+\Delta t)-U_i(t)}$.

It is necessary and sufficient to consider four regions as shown in Fig. 12-3 :

Region 1 : $U_i(t) \geq$ UTP and $V_i(t)=1$

Region 2 : LTP< $U_i(t)$ < UTP and $V_i(t)=1$

Region 3 : LTP< $U_i(t)$ < UTP and $V_i(t)=0$

Region 4 : $U_i(t)$ ≤LTP and $V_i(t)=0$

In region 1 : We must consider four possible cases for $U_i(t+Dt)$

(a) $U_i(t+Dt)>U_i(t)$

(b) LTP<U_i(t+Dt)<U_i(t)

(c) U_i(t+Dt) $\leq$ LTP

(d) U_i(t+Dt)=U_i(t)

In (a) and (b), $V_i(t+Dt)=V_i(t)=1 \Rightarrow \frac{\Delta V_i}{\Delta U_i} = 0$. Therefore $\frac{\Delta E}{\Delta t}=0$.

In (d), $\frac{\Delta U_i}{\Delta t}=0 \Rightarrow \frac{\Delta E}{\Delta t}=0$.

In (c), $V_i(t+Dt)=0 \Rightarrow \frac{\Delta V_i}{\Delta U_i} = \frac{1}{\text{positive number}}$ and $\frac{\Delta U_i}{\Delta t}<0$. Therefore, $\frac{\Delta E}{\Delta t}<0$.

It is concluded that $\frac{\Delta E}{\Delta t} \leq 0$ is always satisfied in region 1.

Similarly, in region (2), region(3), and region (4), $\frac{\Delta E}{\Delta t} \leq 0$ is always satisfied. This completes the proof. **Q.E.D.**

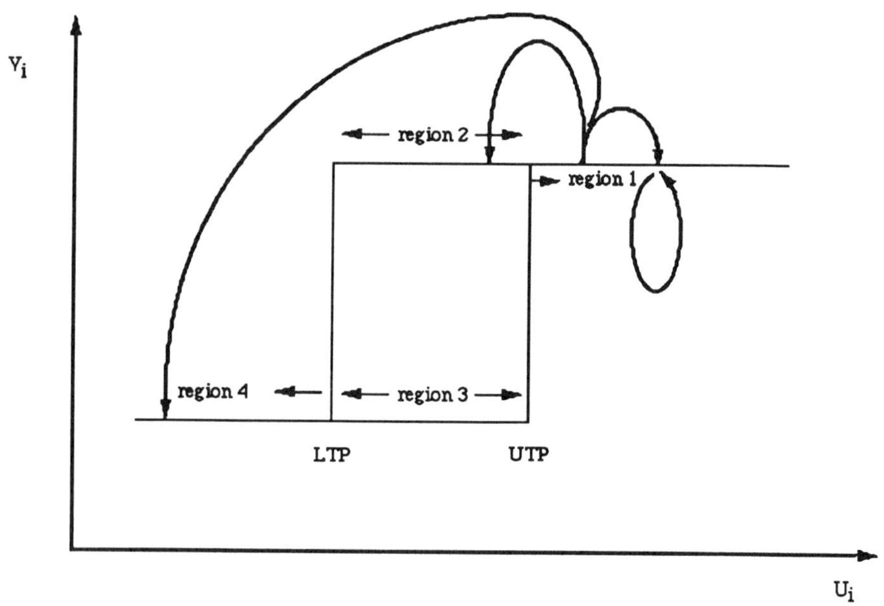

Fig. 12-3 Four possible regions in the hysteresis McCulloch-Pitts neuron

12.7 MAXIMUM NEURAL NETWORK CONVERGENCE

In Theorem 6, the local minimum convergence of the maximum neural network is given.

Theorem 6:

$\frac{dE}{dt} \leq 0$ is satisfied under two conditions such as

(1) $\frac{dU_{m,i}}{dt} = -\frac{\partial E}{\partial V_{m,i}}$ and

(2)

$V_{m,i} = 1$ if $U_{m,i} = \max\{U_{m,1}, U_{m,2}, U_{m,3}, U_{m,4}\}$ and $U_{m,i} \geq U_{m,j}$ for $i > j$

0 otherwise

Proof:

Consider the derivatives of the computational energy E with respect to time t.

$$\frac{dE}{dt} = \sum_m \sum_i \frac{dU_{m,i}}{dt} \frac{dV_{m,i}}{dU_{m,i}} \frac{\partial E}{\partial V_{m,i}}$$

$$= -\sum_m \sum_i \left(\frac{dU_{m,i}}{dt}\right)^2 \frac{dV_{m,i}}{dU_{m,i}} \quad \text{where } \frac{\partial E}{\partial V_{m,i}} \text{ is replaced by } -\frac{dU_{m,i}}{dt}$$

(condition 1)

Let $\frac{dU_{m,i}}{dt}$ be $\frac{U_{m,i}(t+dt)-U_{m,i}(t)}{dt}$. Let $\frac{dV_{m,i}}{dU_{m,i}}$ be $\frac{V_{m,i}(t+dt)-V_{m,i}(t)}{U_{m,i}(t+dt)-U_{m,i}(t)}$.

Let us consider the term $\sum_i \left(\frac{dU_{m,i}}{dt}\right)^2 \frac{dV_{m,i}}{dU_{m,i}}$ for each module separately.

Let $U_{m,a}(t+dt)$ be the maximum at time t+dt and $U_{m,b}(t)$ be the maximum at time t for the module m.

$U_{m,a}(t+dt) = \max\{U_{m,1}(t+dt), U_{m,2}(t+dt), U_{m,3}(t+dt), U_{m,4}(t+dt)\}$

$U_{m,b}(t) = \max\{U_{m,1}(t), U_{m,2}(t), U_{m,3}(t), U_{m,4}(t)\}$

It is necessary and sufficient to consider the following two cases :

1) $a = b$

2) $a \neq b$

If the condition 1) is satisfied, then there is no state change for the module m. Consequently, $\sum_i (\frac{dU_{m,i}}{dt})^2 \frac{dV_{m,i}}{dU_{m,i}}$ must be zero.

If 2) is satisfied, then

$$\sum_i (\frac{dU_{m,i}}{dt})^2 \frac{dV_{m,i}}{dU_{m,i}} = (\frac{U_{m,a}(t+dt)-U_{m,a}(t)}{dt})^2 \frac{V_{m,a}(t+dt)-V_{m,a}(t)}{U_{m,a}(t+dt)-U_{m,a}(t)}$$
$$+(\frac{U_{m,b}(t+dt)-U_{m,b}(t)}{dt})^2 \frac{V_{m,b}(t+dt)-V_{m,b}(t)}{U_{m,b}(t+dt)-U_{m,b}(t)}$$

$$= (\frac{U_{m,a}(t+dt)-U_{m,a}(t)}{dt})^2 \frac{1}{U_{m,a}(t+dt)-U_{m,a}(t)}$$
$$+(\frac{U_{m,b}(t+dt)-U_{m,b}(t)}{dt})^2 \frac{-1}{U_{m,b}(t+dt)-U_{m,b}(t)}$$

$$= \frac{U_{m,a}(t+dt)-U_{m,a}(t)}{(dt)^2} - \frac{U_{m,b}(t+dt)-U_{m,b}(t)}{(dt)^2}$$
$$= \frac{1}{(dt)^2}\{ U_{m,a}(t+dt) - U_{m,a}(t) - U_{m,b}(t+dt) + U_{m,b}(t) \}$$
$$= \frac{1}{(dt)^2}\{ U_{m,a}(t+dt) - U_{m,b}(t+dt) + U_{m,b}(t) - U_{m,a}(t) \}$$
$$> 0$$

because $U_{m,a}(t+dt)$ is the maximum at time t+dt and $U_{m,b}(t)$ is the maximum at time t for the module m.

The contribution from each term is either 0 or positive, therefore

$$\sum_i (\frac{dU_{m,i}}{dt})^2 \frac{dV_{m,i}}{dU_{m,i}} \geq 0 \text{ and } -\sum_m \sum_i (\frac{dU_{m,i}}{dt})^2 \frac{dV_{m,i}}{dU_{m,i}} \leq 0 \Rightarrow \frac{dE}{dt} \leq 0$$

Q.E.D.

12.8 OTHER NEURON MODELS

Hinton et al. introduced a Boltzmann neuron model (Hinton et al. 1984). It is composed of two key elements; a uniform random number generator and an acceptance function. A random number is in a range from zero to one. The acceptance function has two variables; the artificial temperature parameter T and the state of the input. The acceptance function is driven by classical one-dimensional random walk, although Hinton did not explain how to obtain the acceptance function. In 1943 S. Chandrasekhar discussed stochastic problems including radom walk published in Reviews of Modern Physics. We simply use the result of his study for obtaining the Boltzmann acceptance function. In one-dimensional random walk, a particle suffers displacements along a straight line in the form of a series of steps of equal length, each step being taken in the forward, or in backward direction with equal probability. The particle suffers n displacement per unit time. The probability that the particle will find itself between x and x+Δx after a time t is given by:

$$W(x, t)\Delta x = \frac{1}{\sqrt{2\pi n t l^2}} \exp\left(- \frac{x^2}{2ntl^2} \right) \Delta x \qquad (12.8)$$

where l is the length of a step. Based on the calculated particle probability, the Boltzmann acceptance function is given by:

$$f(\sigma, U_i) = \int_0^\infty \frac{1}{\sqrt{2\pi\sigma^2}} \exp\left(- \frac{(x-U_i)^2}{2\sigma^2} \right) dx \qquad (12.9)$$

where σ is $\sigma^2=ntl^2$. If σ=1.70T is assumed, the approximate Boltzmann acceptance function will be given by:

$$f(T, U_i)= \frac{1}{1+e^{-\frac{U_i}{T}}}. \qquad (12.10)$$

When the temperature is very high, the acceptance function is constantly about one half. When the temperature is very low, the Boltzmann neuron behaves like the McCulloch-Pitts binary neuron. If the random number is smaller than the value of the acceptance function, the Boltzmann neuron generates one and zero otherwise. The initial temperature is set to be very high and it is slowly reduced. The method using the temperature scheduling is called simulated annealing. In 1983 Kirkpatrick introduced the first temperature scheduling (Kirkpatrick et al. 1983). Geman and Geman proved the necessary conditions of the global minimum convergence with the method (Geman and Geman 1984).

Szu and Takefuji proposed a Cauchy neuron model (Takefuji and Szu 1989). The difference between the Boltzmann neuron and the Cauchy neuron lies in that the Boltzmann acceptance function is simply replaced by the Cauchy acceptance function. The Cauchy acceptance function is given by:

$$f(U_i, T) = \int_0^{\infty} \frac{1}{\pi T}\left[\frac{dx}{1+(\frac{x-U_i}{T})^2}\right] = \frac{1}{2} + \frac{1}{\pi}\arctan\left(\frac{U_i}{T}\right). \qquad (12.11)$$

Cauchy distribution yields a better chance to converge to neighborhood minima than Gaussian white noise distribution. Cauchy color noise produces both global random flights (long jumps) and local random walks, while Gaussian white noise produces only local random walks (Szu and Hartley 1987). The intensive study on the performance of the Gaussian machine and the Cauchy machine can be found in (Chen 1991).

12.9 REFERENCES

Geman S. and Geman D., (Nov. 1984), Stochastic relaxation, Gibbs distribution and Baysian restoration in images, IEEE Trans. on Patt. Anal. Mach. Int.,vol. PAMI-6.

Hinton G. E., T. J. Sejnowski, and D. H. Ackley, (May 1984), Boltzmann Machines: Constrained Satisfaction Networks that Learn, CMU-CS-84-119, Carnegie Mellon University.

Kirkpatrick S., C.D. Gelatt Jr., and M.P. Vecchi, (May 1983), Optimization by Simulated Annealing, Science, 220, 4598.

Szu H.and Hartley R., (June 1987), Fast Simulated Annealing, Phys. lett. A, 122,

157-162.

Takefuji Y., and Szu H., (1989), Design of parallel distributed Cauchy machines, Proc. of the Int. Joint Conf. on Neural Networks.

Takefuji Y., and Lee K. C., (1991), Artificial neural networks for four-coloring problems and k-colorability problems, IEEE Trans. on Circuits and Systems, 38, 3, pp.326-333.

Takefuji Y. and Lee K. C, (1990), A super parallel sorting algorithm based on neural networks, IEEE Trans. on Circuits and Systems, 37, 11, 1425-1429.

Lee K. C., Takefuji Y., and Funabiki N., (1991), A maximum neural network for the max cut problem, Proc. of International Joint Conference on Neural Networks (IJCNN91-Seattle).

Chapter 13

FORTHCOMING APPLICATIONS

13.1 INTRODUCTION

We have discussed the crossbar switch scheduling problem in Chapter 2 where the neural network representation for N-queen problems in Chapter 1 is modified. A variety of other communication problems are solvable using the similar artificial neural network. For example, time slot assignment in TDM hierarchical switching system and broadcast scheduling in packet radio networks are all similar to the crossbar switch scheduling problem in Chapter 2.

Graph planarization and channel routing problems for VLSI computer aided design are discussed in Chapter 4 and 5. The module orientation problem is introduced as another computer aided design application.

The maximum clique problem are finally introduced to enable the reader to experience the hard graph problem.

13.2 TIME SLOT ASSIGNMENT IN TDM SWITCHING SYSTEMS

TDM (Time Division Multiplexing) hierarchical switching systems have been widely used in terrestrial networks and satellite networks because the system makes it possible to reduce the hardware cost, to enhance the trunk line efficiency, and to accommodate the additional loads (Vaughan 1979) (Joel 1979) (Eng and Acampora

1987) (Bonuccelli 1989). In (Bonuccelli 1989) Bonuccelli defined the general structure of the TDM hierarchical switching systems. As shown in Fig. 13-1, a TDM hierarchical switching system consists of three stages: multiplexers, a central time multiplex crossbar switch, and demultiplexers. Each multiplexer concentrates several input lines from end users to a few input TDM lines to the central time multiplex crossbar switch. The central switch connects each input TDM line with at most one output TDM line. Each demultiplexer connects the output TDM lines with the corresponding output lines to end users. The total number of input lines is equal to that of output lines. Through the TDM hierarchical switching system, each input line can be simultaneously connected with at most one output line and each output line can be also simultaneously connected with at most one input line at any time. Besides, if the i-th multiplexer (demultiplexer) has P_i (Q_i) input (output) lines and I_i (O_i) input (output) TDM lines, at most I_i (O_i) users among the P_i (Q_i) users connected to the i-th multiplexer (demultiplexer) can be simultaneously connected to the central switch.

We can have the same assumption given by Eng's paper (Eng and Acampora 1987). A signal through the switching system is assumed to be composed of fixed-length message packets. A time slot has a unit time where one packet can be transmitted through one line, and it represents one switching configuration. Packets are transmitted in repetitive frames composed of m time slots. The packet transmission between users are performed through multiplexers, the central switch, and demultiplexers where the operations are synchronized. The interconnecting patterns are usually repetitive in every frame but change from slot to slot within a frame.

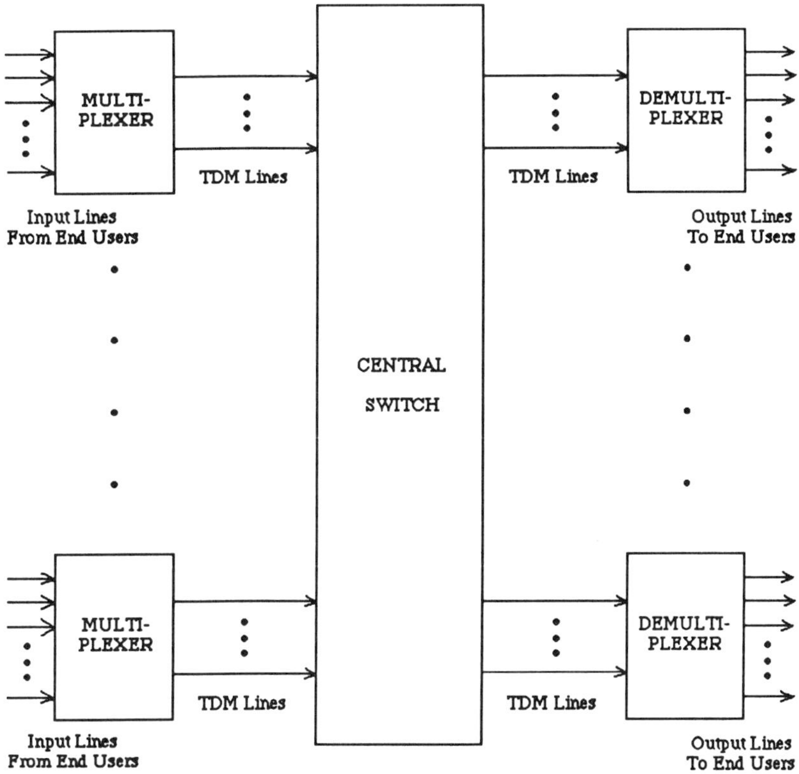

Fig. 13-1 TDM hierarchical switching system

As Bonuccelli defined in his paper (Bonuccelli 1989), a request for the interconnecting pattern configuration through an n×n TDM hierarchical switching system is described by an n×n traffic matrix T where n input lines are connected with n output lines. In the traffic matrix T, each element t_{ij} represents the number of the packets which are requested to be transmitted from the input line #i to the output line #j. t_{ij} is also equal to the number of time slots to transmit the packets. For example $t_{ij}=0$ means that there is no packet and $t_{ij}=3$ means that there are three packets to be transmitted from the input line #i to the output line #j respectively. In order to maximize the throughput of the system, we must find the conflict-free assignment of a given traffic request in the minimum number of time slots.

Fig. 13-2a shows an example of a 4×4 traffic matrix. Fig. 13-2b shows the corresponding 4×4 TDM hierarchical switching system. The traffic matrix shows that there are two packets to be transmitted from the input line #1 to the output line #1, one packet from the input line #2 to the output line #3, one packet from the input line #3 to the output line #4, and one packet from the input line #4 to the output line #1 respectively. The TDM hierarchical switching system consists of two multiplexers, a 2×2 central time multiplex crossbar switch, and two demultiplexers. Each multiplexer has two input lines and it can simultaneously connect at most one of them with the central switch. The central switch has two input TDM lines and two output TDM lines and it can simultaneously connect each input with at most one output and each output with at most one input. Each demultiplexer has two output lines and it can simultaneously connect at most one of them with the central switch.

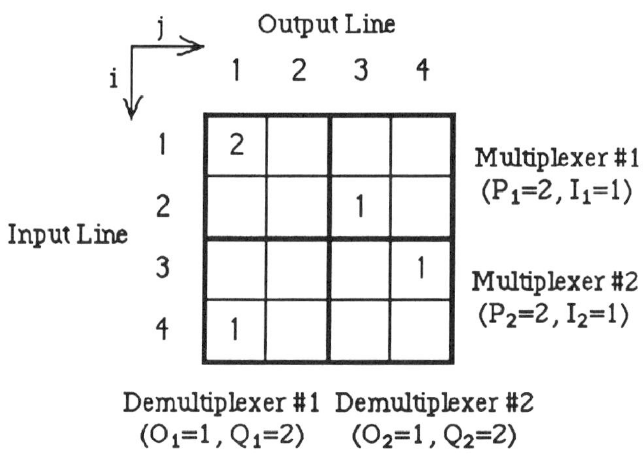

Fig. 13-2 (a) 4 x 4 traffic matrix

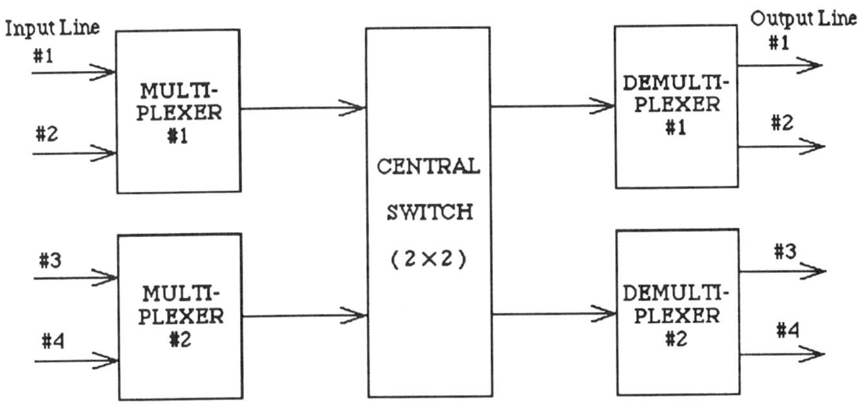

Fig. 13-2 (b) 4 x 4 TDM hierarchical switching system

The problem of finding a conflict-free time slot assignment for the given traffic matrix has been widely studied in the TDM hierarchical switching systems (Eng and Acampora 1987) (Bonuccelli 1989) and in the TDM non-hierarchical switching systems (Inukai 1979) (Inukai 1981) (Schmidt 1969) (Reudink 1978) (Bongiovanni et al. 1981a) (Bongiovanni et al. 1981b) (Gopal et al. 1982) (Lewandowski et al. 1983) (Gopal et al. 1983) (Gopal et al. 1985) (Bertossi et al. 1987), particularly in the TDMA (Time Division Multiple Access) satellite network applications. Although the time slot assignment problem in the TDM hierarchical switching system is not known whether it is NP-complete or not, some problems in the non-hierarchical switching systems are proved to be NP-complete. In 1983 Gopal et al. proved that the traffic scheduling problem in the case of zone interference is NP-complete (Gopal et al. 1983). In 1985 Gopal and Wong proved that the problem of minimizing the number of switching reconfigurations is NP-complete (Gopal et al. 1985). In 1987 Bertossi et al. proved that the problem of finding a time slot assignment which minimizes the total transmission time is NP-complete (Bertossi et al. 1987). The non-hierarchical switching system can be seen as a special case of the hierarchical switching system because the former system does not consider the existence of multiplexers and demultiplexers. In 1987 Eng and Acampora showed the four necessary and sufficient conditions to have valid solutions in the time slot assignment problems in the TDM hierarchical switching systems (Eng and Acampora 1987). In 1989 Bonuccelli proposed the $O(n^5)$ sequential algorithm to find the time slot assignments in the $n \times n$ TDM hierarchical switching systems (Bonuccelli 1989). In 1989 Rose proposed the O(n) parallel algorithm for the simple $n \times n$ crossbar switching system where it is based on a cellular automaton and n^2 processing elements are required (Rose 1989). Rose's algorithm cannot deal with the multiplexer/demultiplexer constraints. Great modifications are required for Rose's algorithm to cope with our time slot assignment problems in the TDM hierarchical switching systems.

References

Bertossi A. A., Bongiovanni G., and Bonuccelli M. A., (June 1987), "Time Slot Assignment in SS/TDMA Systems with Intersatellite Links," IEEE Trans. Commun., Vol. COM-35, pp.602-608.

Bongiovanni G., Tang D. T., and Wong C. K., (July 1981a), "A General Multibeam Satellite Switching Algorithm," IEEE Trans. Commun., Vol. COM-29, pp.1025-1036.

Bongiovanni G., Coppersmith D., and Wong C. K., (May 1981b), " An Optimum Time Slot Assignment Algorithm for an SS/TDMA System with Variable Number of Transponders," IEEE Trans. Commun., Vol. COM-29, pp.721-726.

Bonuccelli M. A., (August 1989), "A Fast Time Slot Assignment Algorithm for TDM Hierarchical Switching Systems," IEEE Trans. Commun., Vol. 37, pp.870-874.

Eng K. Y. and Acampora A. S., (July 1987), "Fundamental Conditions Governing TDM Switching Assignments in Terrestrial and Satellite Networks," IEEE Trans. Commun.,Vol. COM-35, pp.755-761.

Gopal I. S., Bongiovanni G., Bonuccelli M. A., Tang D. T., and Wong C. K., (Nov. 1982), "An Optimal Switching Algorithm for Multibeam Satellite Systems with Variable Bandwidth Beams," IEEE Trans. Commun., Vol. COM-30, pp.2475-

2481.

Gopal I. S., Bonuccelli M. A., and Wong C. K., (Aug. 1983), "Scheduling in Multibeam Satellites with Interfering Zones," IEEE Trans. Commun. Vol. COM-31, pp.941-951.

Gopal I. S., and Wong C. K., (June 1985), "Minimizing the Number of Switchings in an SS/TDMA System," IEEE Trans. Commun., Vol. COM-33, pp.497-500.

Inukai T., (Oct. 1979), "An Efficient SS/TDMA Time Slot Assignment Algorithm," IEEE Trans. Commun., Vol. COM-27, pp.1449-1455.

Inukai T., (1981), "SS-TDMA Networking Via ISL," in Proc. ICC, 1981, pp.70.6.1-70.6.5.

Joel, Jr. A. E., (July 1979), "Digital Switching - How It Has Developed," IEEE Trans. Commun.,Vol. COM-27, pp.948-959.

Lewandowski J. L.,Liu J. W. S., and Liu C. L., (Jan. 1983), "SS/TDMA Time Slot Assignment with Restricted Switching Modes," IEEE Trans. Commun., Vol. COM-31, pp.149-154.

Reudink D. O., (Sept. 1978), "Spot beams promise satellite communication breakthrough," IEEE Spectrum, pp.36-42.

Rose C., (May 1989), "Rapid Optimal Scheduling for Time-Multiplex Switches Using a Cellular Automaton," IEEE Trans. Commun., Vol. 37, pp. 500-509.

Schmidt W. G., (1969), "An On-Board Switched Multiple-Access System for Millimeter-Wave Satellites," 1969 INTELSAT/IEE Int. Conf. Digital Satellite Communication, pp.399-407.

Vaughan H. E., (July 1979), "Research Model for Time-Separation Integrated Communication," IEEE Trans. Commun.,Vol. COM-27, pp.940-947.

13.3 BROADCAST SCHEDULING IN PACKET RADIO NETWORKS

A packet radio network provides data communication services among a large number of geographically distributed and possibly mobile users or nodes through a shared radio channel. The network provides local distribution of information over a wide geographic area. For these distributed and mobile nodes, the network specifications such as the traffic characteristics of the nodes and the network topological structure are often changed. The initial design of the network may be far from the optimal after the network specifications are changed. The packet radio network uses radio as broadcast medium to interconnect the nodes so that the system offers flexibility to the topological changes and has the property that the network capacities can be changed to reflect alterations in the specifications of the network (Roberts 1974) (Gitman et al. 1976) (Kahn 1977) (Kahn et al. 1978) (Kleinrock and Scholl 1980) (Mannel 1980) (Hauptschein and Kajor 1981) (Gafni and Bertsekas 1981) (Segall 1983) (Humblet 1983) (Silvester and Kleinrock 1983) (Shindo et al. 1983) (Tobagi 1984) (Hahn and D. M. Stolle 1984).

In order to provide the communication services among many nodes through a shared radio channel, time division multiple access (TDMA) technology has been used. In TDMA technology the network channel is shared by scheduling each data packet transmission in a short nonoverlapping interval. The nonoverlapping interval is called a time slot which has unit time to transmit one data packet per one node. All the nodes in the network must be synchronized to each time slot.

The packet radio network has the following constraints. 1) The network has symmetric connectivities between nodes. In other words, if node #i can receive a data packet from node #j, then node #j can also receive a data packet from node #i. These two nodes are called neighbors. 2) A transmitted data packet may be received by all the neighboring nodes of the sending node. 3) A node cannot send and receive a data packet simultaneously. 4) A node cannot receive more than one data packets simultaneously. If two or more nodes transmit data packets to a node simultaneously, the node cannot receive any of the data packets successfully. Fig. 13-3a shows an example of the packet radio network. The edges between the vertices or nodes represent the connectivities between the nodes which mean that the two nodes are neighbors. For example, node #1 and node #3 are the neighbors of node #2. When node #2 sends a data packet in a time slot, node #1 and node #3 can receive the data packet. During this time slot, node #2 cannot receive a data packet, and node #1 and node #3 cannot send data packets, either. If node #4 also sends a data packet in this time slot, a conflict occurs at node #3 and node #3 cannot receive any data packet successfully.

Fig. 13-3 (a) Packet radio network with 7 nodes

As shown in Fig. 13-3b, the network connectivities are described in an n×n symmetric binary matrix where n is the number of nodes. We call it the network matrix. The black squares in Fig. 13-3b indicate that the corresponding two nodes are neighbors, and the white squares indicate that the two nodes are not neighbors.

Fig. 13-3 (b) Adjacency matrix of the network in Fig. 13-3a

The problem of finding a conflict-free broadcasting schedule in the given radio network has been extensively studied and several sequential algorithms have been proposed (Baker and Ephremides 1981) (Baker et al. 1982) (Baker et al. 1984) (Arikan 1984) (Even et al. 1984) (Nelson and Kleinrock 1985) (Chlamtac and Kutten 1985) (Post et al. 1985) (Chlamtac and Lerner 1985) (Robertazzi and Sarachik 1986)

(Chlamtac and Pinter 1987) (Chlamtac and Lerner 1987) (Hajek and Sasaki 1988) (Cidon and Sidi 1989) (Ephremides and Truong 1990). Baker et al. first formulated the problem and proposed some ad hoc algorithms for it (Baker and Ephremides 1981) (Baker et al. 1982) (Baker et al. 1984). The cycle in which all the nodes in a given network can send data packets without causing any conflict is called TDMA cycle. The problem of finding TDMA cycle has been most intensively studied. Even et al. proved that the problem of minimizing the length of TDMA cycle is NP-complete (Even et al. 1984). Recently two sequential algorithms have been proposed for this problem. In 1989 Cidon et al. proposed two algorithms. One is round-robin algorithm where the priorities of the nodes are cycled. The other is wait-for-neighbors algorithm where the assigned nodes are eliminated from the following assignment procedure before all the neighboring nodes are assigned (Cidon and Sidi 1989). In 1990 Ephremides et al. proposed an algorithm where first each node is assigned at each time slot diagonally and then the nodes with higher priorities are assigned in the remaining rooms of time slots (Ephremides and Truong 1990). However, these algorithms do not provide optimization for minimizing the length of TDMA cycle.

Although parallel algorithms for the broadcast scheduling problems have been highly demanded since the advent of the inexpensive VLSI technology, few parallel algorithms have been reported.

References

Arikan E., (July 1984), "Some Complexity Results about Packet Radio Networks," IEEE Trans. Inform. Theory, Vol. IT-30, 4, pp.681-685.

Baker D. J.and Ephremides A., (Nov. 1981), "The Architectual Organization of a Mobile Radio Network via a Distributed Algorithm," IEEE Trans. Commun., Vol. COM-29, 11, pp.1694-1701.

Baker D. J., Ephremides A., and Flynn J. A., (Jan. 1984), "The Design and Simulation of a Mobile Radio Network with Distributed Control," IEEE J. Select. Areas Commun., Vol. SAC-2, 1, pp.226-237.

Baker D. J., Wieselthier J., and Ephremides A., (1982), "A Distributed Algorithm for Scheduling the Activation of Links in a Self-Organizing, Mobile, Radio Network,"

Proc. IEEE Int. Conf. Commun., pp.2F.6.1-2F.6.5.

Chlamtac I.and Kutten S., (Dec. 1985), "On Broadcasting in Radio Networks - Problem Analysis and Protocol Design," IEEE Trans. Commun., Vol. COM-33, 12, pp.1240-1246.

Chlamtac I.and Lerner A., (1985), "A Link Allocation Protocol for Mobile Multi-hop Radio Networks," Proc. GLOBECOM '85, pp.238-242.

Chlamtac I.and Lerner A., (July 1987), "Fair Algorithms for Maximal Link Activation in Multihop Radio Networks," IEEE Trans. Commun., Vol. COM-35, 7, pp.739-746.

Chlamtac I. and Pinter S. S., (June 1987), "Distributed Nodes Organization Algorithm for Channel Access in a Multihop Dynamic Radio Network," IEEE Trans. Computers, Vol. C-36, 6, pp.728-737.

Cidon I. and Sidi M., (Oct. 1989), "Distributed Assignment Algorithms for Multihop Packet Radio Networks," IEEE Trans. Computers, Vol. 38, 10, pp.1353-1361.

Ephremides A. and Truong T. V., (April 1990), "Scheduling Broadcasts in Multihop Radio Networks," IEEE Trans. Commun., Vol. 38, 4, pp.456-460.

Even S. et al., (1984), "On the NP-Completeness of Certain Network Testing Problems," Networks, Vol. 14, pp.1-24.

Gafni E. M.and Bertsekas D. P., (Jan. 1981), "Distributed Algorithms for Generating Loop-Free Routes in Networks with Frequently Changing Topology," IEEE Trans. Commun., Vol. COM-29, 1, pp.11-18, Jan. 1981.

Gitman I, Vanslyke R. M., and Frank H., (Aug. 1976), "Routing in Packet-Switching Broadcast Radio Networks," IEEE Trans. Commun., Vol. COM-24, pp.926-930.

Hahn J. J. and Stolle D. M., (Nov. 1984), "Packet Radio Network Routing Algorithms: A Survey," IEEE Communications, Vol. 22, 11, pp.41-47.

Hajek B. and Sasaki G., (Sept. 1988), "Link Scheduling in Polynomial Time," IEEE Trans. Inform. Theory, Vol. 34, 5, pp.910-917.

Hauptschein A. and Kajor M., (July 1981), "Recognition and Self-Organization of Nodes into DTDMA Nets," IEEE Trans. Aerosp. Electron. Systems, Vol. AES-17,

4, pp.531-542.

Humblet P. A., (June 1983), "A Distributed Algorithm for Minimum Weight Directed Spanning Trees," IEEE Trans. Commun., Vol. COM-31, 6, pp.756-762.

Kahn R. E., (Jan. 1977), "The Organization of Computer Resources into a Packet Radio Network," IEEE Trans. Commun., Vol. COM-25, 1, pp.169-178.

KahnR. E. et al., (Nov. 1978), "Advances in Packet Radio Technology," Proc. IEEE, Vol. 66, 11, pp.1468-1496.

Kleinrock L.and Scholl M. O., (July 1980), "Packet Switching in Radio Channels: New Conflict-Free Multiple Access Schemes," IEEE Trans. Commun., Vol. COM-28, 7, pp.1015-1029.

Mannel W. M., (Sept. 1980), "Future Communications Concepts in Support of U.S. Army Command and Control," IEEE Trans. Commun., Vol. COM-28, 9, pp.1540-1550.

Nelson R. and Kleinrock L., (Sept. 1985), "Spatial TDMA: A Collision-Free Multihop Channel Access Protocol," IEEE Trans. Commun., Vol. COM-33, 9, pp.934-944.

Post M. J., Kershenbaum A. S., and Sarachik P. E., (1985), "A Distributed Evolutionary Algorithm for Reorganizing Network Communications," in Proc. IEEE MILCOM Conf., pp.133-139.

Robertazzi T. G. and Sarachik P. E., (Jan. 1986), "Self-Organizing Communication Networks," IEEE Communications, Vol. 24, 1, pp.28-33.

Roberts L. G, (Feb. 1974), "Data by the Packet," IEEE Spectrum, pp.46-51.

Segall A., (Jan. 1983), "Distributed Network Protocols," IEEE Trans. Inform. Theory, Vol. IT-29, Shindo S. et al., (1983), "TDMA for Radio Local Distribution System," Proc. IEEE Int. Commun. Conf., pp.370-374.

Silvester J. A.and Kleinrock L., (Aug. 1983), "On the Capacity of Multihop Slotted ALOHA Networks with Regular Structure," IEEE Trans. Commun., Vol. COM-31, 8, pp.974-982.

Tobagi F. A., Binder R., and Leiner B., (Nov. 1984), "Packet Radio and Satellite Networks," IEEE Communications, Vol. 22, 11, pp.24-40.

13.4 MODULE ORIENTATION PROBLEMS

Since Cook presented the foundations for the theory of NP-completeness in 1971 (Cook 1971), a large number of commonly encountered problems from mathematics, computer science, molecular biology, management science, seismology, communications, and operations research have been known to be NP-complete. Garey and Johnson have collected more than 300 NP-complete problems (Garey and Johnson 1979). It is widely believed that no polynomial time algorithm exists in finding the optimum solution of any NP-complete problems (Garey and Johnson 1979). Although there is no known polynomial time algorithm for solving the NP-complete problem, a polynomial time algorithm may exist on a particular instance of the problem. An algorithm for NP-complete problems must be carefully verified based on time complexity or tested by massive instances (examples) of the problem using empirical simulation runs. To practically cope with NP-complete problems, we have been forced to be satisfied with finding a near-optimum or a good solution within an acceptable time instead of finding an optimum solution.

The NP-completeness of the module orientation problem was proven by Liu (Ran Libeskind-Hadas and Liu 1989). For physical design of very large scale integrated circuits, it is important to properly place the given modules under certain constraints such as the minimum area and/or the minimum total wire length. The module orientation problem was first proposed by Libeskind-Hadas and Liu (Ran Libeskind-Hadas and Liu 1989). The first assumption of the problem is that all the modules have already been placed according to some placement algorithm (Hanan et al. 1976) (Preas and Karger 1988). The second assumption is that the pin positions on each module are fixed. The goal of this problem is to minimize the total wire length by flipping each module with respect to its vertical and/or horizontal axes of symmetry as shown in Fig. 13-4. This means there are only four possible orientations for each module. Libeskind-Hadas and Liu called the problem as module orientation problem : finding the optimal flips for a given set of modules. Another similar problem was discussed at the same time by Libeskind-Hadas and Liu : module rotation problem. Libeskind-Hadas and Liu have proved the module orientation problem and module rotation problem are both NP-Complete (Ran Libeskind-Hadas and Liu 1989).

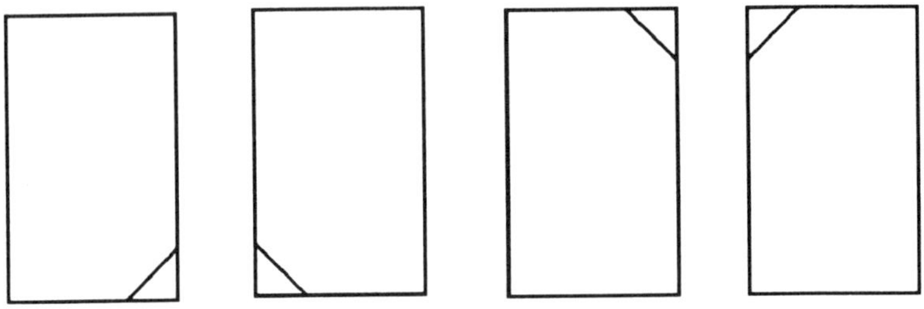

Fig. 13-4 Four orientations for one module

For simplicity, the Euclidean metric is used to define the distance between two pins in the module orientation problem. Let p and q be two pins which belong to the same net N. Let (x_p,y_p) and (x_q,y_q) represent the positions of p and q according to a certain orientation of the modules. The goal is to minimize the total wire length where the summation is carried out over all pairs of pins which belong to the same net. That is to minimize L where L is given by

$$L = \sum_{p,\,q \in N} \sqrt{(x_p-x_q)^2+(y_p-y_q)^2} \tag{13.1}$$

The problem is to determine the optimum orientation for each module in order to minimize L. Libeskind-Hadas and Liu mapped the problem into the Hopfield neural network by using the 4 x M neural array where M is the number of modules and each module has 4 possible orientations.

References

Cook S. A, Proc. of 3rd Ann. ACM Symp. on Theory of Computing (ACM, New York, 1971), pp. 151-158.

Garey M. R. and Johnson D. S., Computers and Intractability : A Guide to the

Theory of NP-Completeness, (W. H. Freeman and Co., San Francisco, 1979).

Ran Libeskind-Hadas and Liu C. L., Proc. 26th Design Automation Conference (ACM, New York, 1989), pp. 400-405.

Hanan M., Wolff Sr. P. K., and Agule B. J., (1976), J. of Design Automation and Fault Tolerant Computing, 1, pp. 28-61.

Preas B. T. and Karger P. G., Physical Design Automation of VLSI Systems, (B. Preas and M. Lorenzetti eds., Menlo Park, California: Benjamin-Cummings, Inc., 1988), Chapter 4, pp. 87-155.

13.5 MAXIMUM CLIQUE PROBLEMS

A graph G is given with a set of vertices and a set of edges. When the graph G has an edge between #i vertex and #j vertex, #i vertex and #j vertex are called adjacent. A clique of the graph G is a complete subgraph of G where any two vertices are adjacent to each other. The maximum clique is a clique which has the largest number of vertices among cliques of G. The problem of finding a maximum clique of an arbitrary graph is known to be NP-complete (Garey and Johnson 1979). This means that in the worst case the computing time for solving the maximum clique problem grows exponentially with the graph size.

Fig. 13-5a shows a graph with eight vertices and 17 edges (Carraghan and Pardalos 1990). For example, #1 vertex and #2 vertex in this graph are adjacent because the graph has an edge between #1 vertex and #2 vertex. It is known that the maximum clique of this graph has four vertices and two maximum cliques, {2, 3, 4, 7} and {2, 4, 6, 7}, exist as shown in Fig. 13-5 (b) and (c).

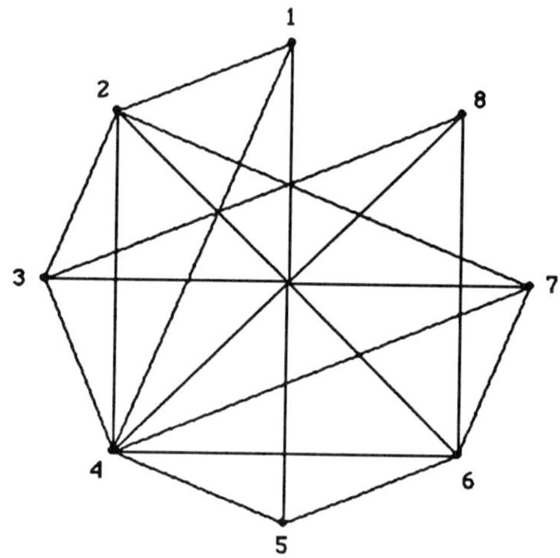

Fig. 13-5 (a) An eight-vertex graph

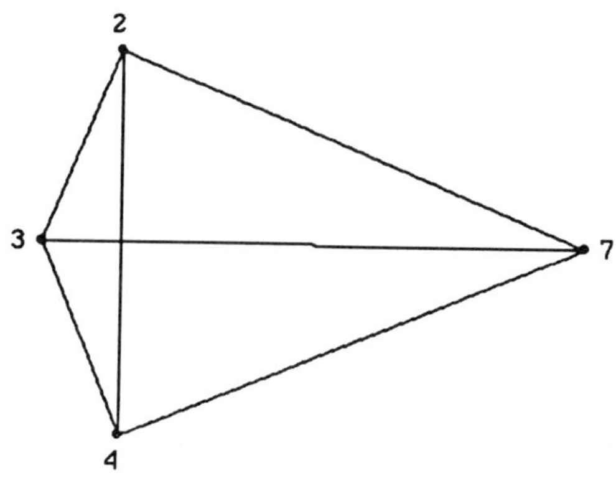

Fig. 13-5 (b) The maximum clique #1

Several sequential algorithms for the maximum clique problem have been reported (Balas and Yu 1986) (Pardalos and Phillips 1990) (Carraghan and Pardalos 1990) (Amin and Hakim 1972) (Bron and Kerbosch 1973) (Gerhards and Lindenberg 1979) (Drake 1986). The computation time of the existing algorithms not only depends on the number of vertices of the graph but also on the number of the edges. In other words, the existing algorithms can neither deal with the large size graphs nor with high density graphs. Note that the graph density means the ratio between the number of edges in the graph with n vertices and that in the complete graph with n vertices.

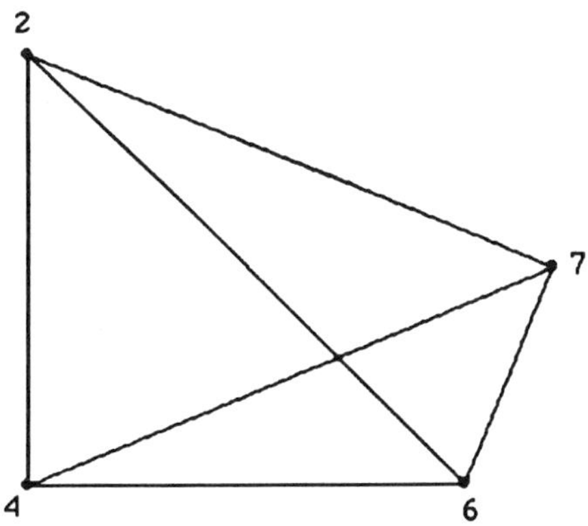

Fig. 13-5 (c) The maximum clique #2

In 1986 Balas and Yu proposed an O(n+m) time branch and bound algorithm where n is the number of vertices and m is the number of edges in the graph. Balas's algorithm is only tested by small size graphs where the largest graph has 400 vertices and 30,000 edges. Their algorithm cannot deal with high density graphs because of prohibitively long computation time. In 1990 Pardalos and Phillips proposed a sequential algorithm where they formulated the maximum clique problem as a linearly constrained indefinite quadratic global optimization problem (Balas and Yu 1986). Although they used the supercomputer Cray 2, they gave up solving larger than 75-vertex graph problems. Their algorithm requires more than one hours on Cray 2 to solve the 75-vertex graph problem with 91% density. In 1990 Carraghan and Pardalos proposed a sequential algorithm based on a partial enumeration. Carraghan's algorithm is able to solve up to 3000-vertex and over 1,000,000-edge graph problems on the main frame IBM 3090. However, Carraghan's algorithm also requires prohibitively long computation time even for a middle size graph problem where it takes more than one hours to solve a 1000-vertex graph problem with 40% density on IBM 3090.

In 1980 Mead and Conway proposed a parallel algorithm for the maximum clique problem where it requires $O(n^2)$ time with maximum $O(2^n)$ processors for an n-vertex graph problem. Few parallel algorithms have been reported within our knowledge.

Several applications of the maximum clique problem for real world problems have been reported. Balas and Yu introduced four applications - information retrieval, experimental design, signal transmission, and computer vision (Balas and Yu 1986). Ogawa proposed an application for the labeled point pattern matching problem (Ogawa 1986). Lecky et al. proposed an application for the PLA folding problem (Lecky 1989). Horaud and Skordas proposed an application for the stereo vision correspondence problem (Horaud and Skordas 1989).

References

Amin A. T. and Hakim S. L., (1972), "Upper bounds on the order of a clique of a graph," SIAM J. Appl. Math., Vol. 22, 4, pp. 569-573.

Balas E. and Yu C. S., (Nov. 1986), "Finding a maximum clique in an arbitrary graph," SIAM J. Comput., Vol. 15, 4, pp. 1054-1068.

Bron C. and Kerbosch J., (1973), "Finding all cliques of an undirected graph," Commun. ACM, Vol. 16, 9, pp. 575-579.

Carraghan and P. M. Pardalos, (Nov. 1990), "An exact algorithm for the maximum clique problem," Operations Research Letters, 9, pp. 375-382.

Drake D. A., (1986), "Embedding maximal clique of sets in maximal cliques of bigger sets," Discrete Math., 58, pp. 229-242.

Garey M. R. and Johnson D. S., Computers and Intractability : A Guide to the Theory of NP-Completeness, (W. H. Freeman and Co., San Francisco, 1979).

Gerhards L. and Lindenberg W., (1979), "Clique detection for nondirected graphs: two new algorithms," Computing, 21, pp. 295-322.

Horaud R. and Skordas T., (1989), "Stereo correspondence through feature grouping and maximal cliques," IEEE Trans. Pattern Analy. Mach. Intelligence, Vol. 11, 11, pp. 1168-1180.

Lecky J. E., Murphy O. J., and Absher R. G., (1989), "Graph theoretic algorithms for the PLA folding problem," IEEE Trans. CAD., Vol. 8, 9, pp. 1014-1021.

Ogawa H., (1986), "Labeled point pattern matching by Delaunay triangulation and maximal cliques," Pattern Recognition, Vol. 19, 1, pp. 35-40.

Pardalos P. M. and Phillips A. T., (1990), "A global optimization approach for solving the maximum clique problem," Int. J. Computer Mathematics, Vol. 33, pp. 209-216.

Chapter 14

CONJUNCTOIDS AND ARTIFICIAL LEARNING

Multinomial Conjunctoids are supervised statistical modules that learn the relationships among binary events. The multinomial conjunctoid algorithm precludes the following problems that occur in the existing feedforward multi-layered neural networks: (a) the existing neural networks often cannot determine underlying neural architectures, for example how many hidden layers should be used, how many neurons in each hidden layer are required, and what interconnections between neurons should be made; (b) the existing neural networks cannot avoid convergence to suboptimal (local minimum) solutions during the learning process; (c) the existing network networks require many iteration steps to converge, if at all, to stable states; and (d) the existing neural networks may not be sufficiently general to reflect all learning situations. By contrast, multinomial conjunctoids are based on a well-developed statistical decision theory framework, which guarantees that learning algorithms will converge to optimal learning states as the number of learning trials increases, and that convergence during each trial will be very fast. In this Chapter a prototype multinomial conjunctoid circuit based on the CMOS VLSI technology is described.

14.1 INTRODUCTION

Artificial neural network models have been studied for many years in the hope of achieving human-like performance in such fields as speech and image recognition (McClelland et al. 1986). A recent resurgence in the field of artificial neural networks has resulted from VLSI advances, new neural network models, and learning algorithms such as competitive learning, Boltzmann learning, and sigma-pi back propagation learning (McClelland et al. 1986) (Hopfield and Tank 1985) (Tank and Hopfield 1986) (Hinton et al. 1984). However even the most prominent learning algorithms, which are based on sigma-pi unit back propagation, have the following serious problems (McClelland et al. 1986): (a) It is usually very difficult to determine a multi-layered neural architecture, for example how many hidden layers should be used, how many neurons in each hidden layer are required, and what interconnections between neurons should be made; (b) there are possibilities of suboptimal convergence during the sigma-pi back propagation process; (c) it is possible that convergence will require prohibitively many iterations during each learning trial; and (d) there is no guarantee that sigma-pi back propagation units are sufficiently general to reflect all learning situations.

In this Chapter a new statistical learning model called the multinomial conjunctoid will be reviewed (Jannarone et al. 1988). Multinomial conjunctoids are based on a well-developed statistical decision theory framework, which guarantees that conjunctoid learning will converge to optimal states over learning trials and the learning will be fast during learing trials. In addition, a simple multinomial conjunctoid module based on CMOS VLSI technology will be introduced.

14.2 MULTINOMIAL CONJUNCTOID CONCEPTS

Conjunctoids are functional parallel distributed processing modules that are based on a certain probability framework. The framework treats each observed (X, Y) combination as a realization of a multivariate (binary) random variable:

$$\underset{1 \times (M+N)}{W} = \underset{1 \times M}{(X,} \ \underset{1 \times N}{Y)} \tag{14.1}$$

where X and Y are M binary input variables and N binary output variables respectively. The probability framework also assumes the existence of specific likelihood functions for samples based on W. These likelihoods include estimable parameters that can be used to both evaluate and utilize (X,Y) associations. When $M+N$ is small and reflecting all possible associations among X and Y are necessary, it is convenient to assume that W has the multinomial likelihood:

$$\Pr\{W = w \mid \alpha\} = \frac{\alpha_w}{\sum_{u \in \beta^{M+N}} \alpha_u}, \quad w \in \beta^{M+N} \tag{14.2}$$

where $\beta^{M+N} = \{w\!: wk = 0, 1;\ k = 1, \ldots, 2^{M+N}\}$ and the parameter vector α satisfies $0 < \alpha_u < 1$ ($u \in \beta^{M+N}$). Multinomial conjunctoid learning occurs during a series of learning trials, when W values are observed and α values are estimated. A useful consequence of the parametric probability framework is that conditional output probabilities for given input values can also be described by estimable parameters. For the multinomial case these probabilities take the form:

$$\Pr\{Y = y \mid X = x\,;\alpha\} = \frac{\alpha_{(x,\,y)}}{\sum_{v \in \beta^{N}} \alpha_{(x,\,v)}} \tag{14.3}$$

Fig. 14-1 illustrates how a multinomial machine learns parity in the case where M = 3, N = 1, and "almost" no Bayes structure is used ("almost", because defining probabilities before the first learning trial requires a weak Bayes prior distribution (Jannarone et al. 1988)). Initially, the conjunctoid assigns a probability of 0.5 to both y = 1 and y = 0 for each possible x value, in lieu of any experience that would point toward the correct y values. This is indicated by the value of 0.5 for the 8 estimated output probability graphs before the first learning trial. The effect of the first trial value, $(x_1, x_2, y) = (0, 0, 0)$, is shown in the estimated output probabilities graph for $x_1=x_2=0$. During the first learning trial the estimate of $\Pr\{Y = 1 \mid X = 0\}$ shifts from 0.5 to 0, whereas the estimated $\Pr\{Y = 0 \mid X = 0\}$ shifts from 0.5 to 1. Other learning trials have similar effects on appropriate y probabilities, as Fig. 14-1 shows. Overall expected proportion of correct guesses will monotonically increase with learning trial.

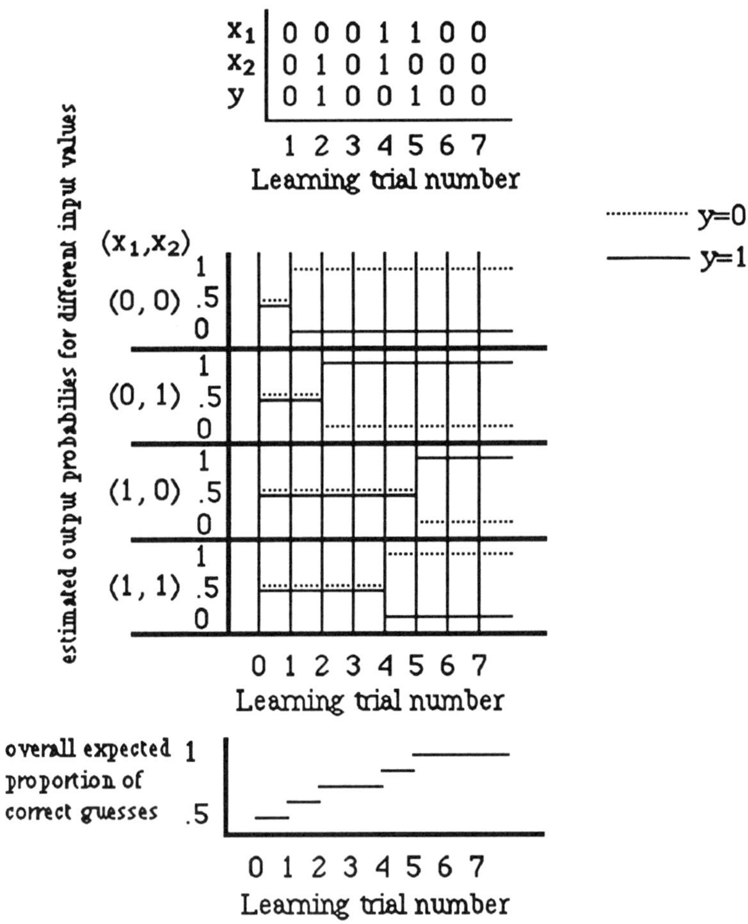

Fig. 14-1 Estimated output (y) probabilies for different input (x_1, x_2) values and overall expected proportion of correct guesses with respect to learning trials

14.3 MULTINOMIAL CONJUNCTOID CIRCUITRY

Fig. 14-2 shows a schematic diagram for a multinomial conjunctoid module. The diagram is composed of interconnections among several functional units, called elementary processing units, that include 2^{M+N} parameter estimators, a parameter multiplexer, 2^N output pattern accumulators, and an output comparator. Multinomial conjunctoid circuitry can also be grouped into larger "experience" and "performance" segments. During each learning cycle the experience segment receives prior/learning data and sends them to the parameter estimators, which in turn send current parameter estimates to the performance segment. At the start of each learning cycle a unit of prior/learning data--consisting of a single (x, y) observation, w, along with a positive learning important weight, L--is passed to each parameter estimator. Next, each parameter estimator--consisting of an input indicator, an estimate updator, and an output register-performs two functions: (a) the input indicator sets a flag, u, to 1 if w matches the parameter for its estimator unit, and to 0 otherwise; and (b) the parameter updator modifies the output register by setting

$$\alpha_{new} = \frac{\alpha_{old}+L\cdot u_w}{1+L}. \qquad (14.4)$$

In line with Bayes developments described in (Jannarone et al. 1988), L corresponds to the "sample size of each new learning trial" relative to the "prior sample size". Before any prior learning trials occur, α values are set to an initial value of .5 for each parameter. Also each parameter estimator performs separately from and simultaneously with all others, so that each learning cycle is very fast. Regarding performance unit functioning, just as the experience segment of Fig. 14-2 executes one learning cycle for each input (w, L) learning unit, the performance segment executes one behavior cycle for each input x value. At the beginning of each behavior cycle, the parameter multiplexer uses the input x value to admit only the 2^N parameter estimates, $\{\ \alpha(x,v),\ v \in \beta^{M+N}\ \}$ associated with the input x value, among the 2^{M+N} parameter estimates coming from the experience segment. Next, each output pattern accumulator selects and stores the single estimate coming from the parameter multiplexer that corresponds to its associated y value. Finally, the output comparator identifies the single output pattern having the highest estimated parameter value (or any one of the highest-valued accumulators in case of a tie) and outputs its y value. Thus, as with the experience segment the performance segment functions quite simply. Moreover, each behavior cycle is quick because the parameter multiplexer performs no computations, the output pattern accumulators function simultaneously, and the output comparator's sole task is to locate the address containing the largest value among 2^N words of storage.

The experience stage is composed of F_1 and F_2 function boxes. Each of F_1 and F_2 box has a unique identifier (ID), w, with a range of 0 to $2^{M+N}-1$. Thus, given an input w value, a single F_1 box having an ID matching the input data value fires and generates $u_w = 1$ and the other F_1 boxes generate $u_w = 0$. In addition F2 boxes compute

$$\alpha_{new} = \frac{\alpha_{old}+L\cdot u_w}{1+L},$$

where L is k-bit. Hence only one F_2 box computes

$$\alpha_{new} = \frac{\alpha_{old}+L}{1+L},$$

and the others compute

$$\alpha_{new} = \frac{\alpha_{old}}{1+L}.$$

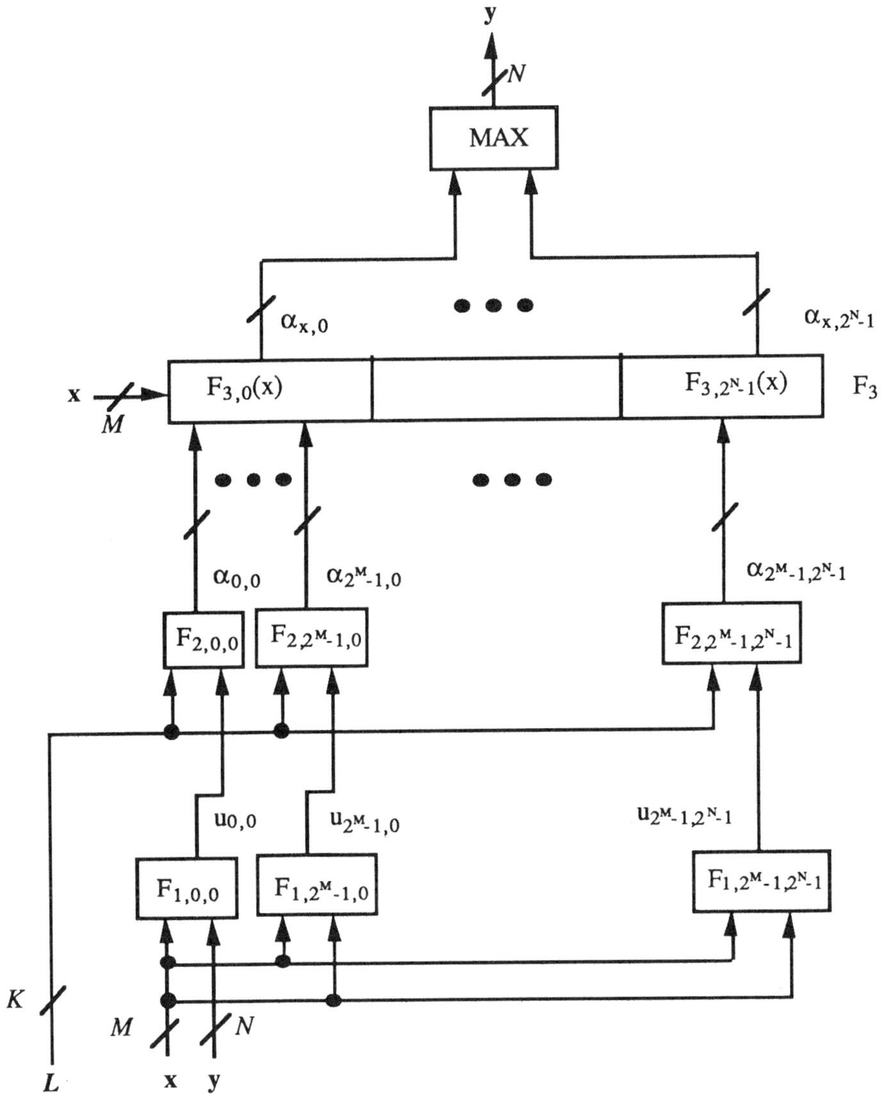

Fig. 14-2 A functional diagram of the multinomial conjunctoid

In the performance stage the F_3 box performs a joint multiplexing function. There are 2^N multiplexers provided in the F_3 box so that the total of 2^N outputs (a values)

are generated. Each multiplexer selects one of 2^M values based on the given x. The MAX function box chooses the biggest among the 2^N a values and generates y, the subscript of the largest ay value. We have developed a simple multinomial conjunctoid module based on digital logic design which has four bits for x and two bits for y in a CMOS chip. We also have implemented analog circuit modules for multinomial conjunctoids. Fig. 14-3 shows the analog F_1 and F_2 circuitry. The analog value corresponding to α_{old} is held in a capacitor and the value is sent through the adder and divider circuits to generate anew. (The voltage follower circuits prevent the capacitor from discharging through the remaining circuitry.) The capacitor's input and output are connected through complementary switches to compute

$$\alpha_{new} = \frac{\alpha_{old}+L\cdot u_w}{1+L}.$$

The numerator of the equation is computed by the inverting amplifier. The analog L value is added to the numerator only when the F_1 box generates u = 1 by manipulating the analog switch. The numerator is divided by the denominator using an analog divider and the result is stored in the capacitor.

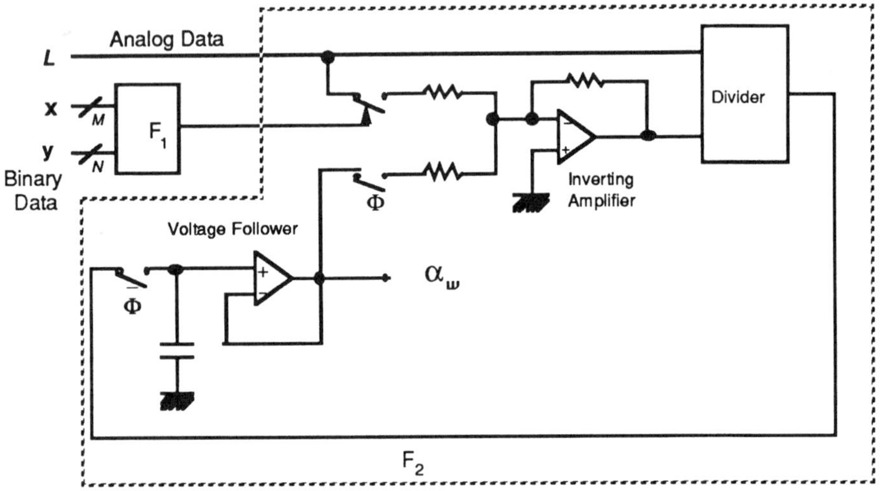

Fig. 14-3 F_1 and F_2 units

In analog multiplexers using analog switches, the given x appropriate α_W values are chosen in each of multiplexers.

We have briefly described the multinomial conjunctoid machine, which is guaranteed to overcome certain problems associated with existing alternatives. We are also investigating the following problems: how to manage the learning rate L (learning scheduling), how much resolution is required for L and α, what sizes of x and y can be dealt with in a single chip, and how to deal with larger problems.

14.4 REFERENCES

Hinton G. E., Sejnowski T. J., and Ackley D. H., (1984), Boltzmann Machines: Constrained Satisfaction Networks that Learn, CMU-CS-84-119, CMU.

Hopfield J. J., and Tank D. W., (1985), Neural Computation of Decisions in Optimization Problems, Biological Cybernetics, 52, 141-152.

Jannarone R. J., Yu K. F., and Takefuji Y., (1988), Conjunctoids: Statistical Learning Modules for Binary Events," Neural Networks, 1, 4.

McClelland J. L., Rummelhart D. E., and the PDP Research Group, (1986), Parallel Distributed Processing Vol.I and Vol.II, MIT Press.

Tank D. W., and Hopfield J. J., (1986), Simple Neural Optimization Networks: An A/D Converter, Signal Decision Circuit, and a Linear Programming Circuit, IEEE Trans. on Circuits and Systems, Vol. CAS-33, No.5, pp.533-541.